BASICS OF OPTICS

BASICS OF OPTICS

Rajnish Pandit

ANMOL PUBLICATIONS PVT. LTD.
NEW DELHI - 110 002 (INDIA)

ANMOL PUBLICATIONS PVT. LTD.
H.O.: 4374/4B, Ansari Road, Darya Ganj,
New Delhi-110 002 (India)
Ph.: 23278000, 23261597
B.O.: No. 1015, Ist Main Road, BSK IIIrd Stage
IIIrd Phase, IIIrd Block,
Bangalore - 560 085 (India)
Visit us at: www.anmolpublications.com

Basics of Optics

ISBN 978-81-261-3556-1

PRINTED IN INDIA

Printed at Mehra Offset Press, Delhi.

Contents

Preface

This book was motivated by my experience in teaching the Physics course in the Electronics Engineering program at the University of Amravati. *Introduction to Optics and Optical Imaging* is designed for students, engineers, and optical scientists seeking a comprehensive background in physical optics in the minimum time with the minimum number of mathematical and physical concepts. Maxwell's equations are taken as the basic foundation of the book, and this set of four vector differential equations is shown to spawn the three major trends in understanding and describing optical phenomena today: ray theory, diffraction integral theory, and plane wave spectrum theory.

All of the material contained in this book is explained by one (and frequently, more than one) of these three approaches. When possible, the interrelationships between these three methods are detailed. In this way, many diverse physical optics phenomena are explained in terms of only three different fundamental viewpoints, which themselves stem from the four Maxwell equations. Since the optical theory is developed from the broader viewpoint of electromagnetic theory (i.e., from Maxwell's equations) rather than from the traditional approach of classical optics, certain seemingly disjointed historical concepts such as "Huygens sources," "Kirchhoff Integrals," "light rays," and "the Abbe plane wave viewpoint" are replaced by the more modern concepts of "Greens functions," "the Stratton-Chu (and vector potential) formulation," "the high frequency (zero wavelength) limit of the diffraction integral approach," and "the plane wave spectrum viewpoint," respectively. In this way, the ideas of classical and modern optics are integrated within a unified framework that uses general electromagnetic theory as its foundation. This provides a broad-based under standing of the subject along with a solid foundation for future study and research. The presentation in this book is based on a combination of intuitive explanations and vector mathematics. The purpose of the text is to explain, via a minimum number of fundamental principles, the concepts being presented.

The purpose of the equations is to provide a basis for definitizing the ideas in the text, as well as for making practical calculations (and, of course, for developing software, if the reader is so inclined). In this way, readers primarily interested in acquiring the concepts of optics and optical imaging can read the text only and skim through the mathematics. Readers interested in a more in depth analytical knowledge of the subject can get that from working through the equations.The dual presentation, math plus exposition, is at the heart of what this book is all about namely, *understanding*. The purpose of this book is to open up the field of optics and optical imaging so that, having read the book, the reader may understand not only modern imaging systems, but also the underlying principles of optics enough to become a developer of future optical technology or software.

Author

Chapter 1

A Brief History of Optics

THE ANCIENT WORLD

The earliest known mirrors are made of polished obsidian (volcanic glass) and date from ~7000 BCE. They were found at Catal Höyük (now in Turkey).

Ancient Egypt, China and Greece all used polished metal mirrors <1000 BCE. There are early references to burning mirrors. Early lenses made of crystal.

In this period, there were innumerable confusions and false starts toward an understanding of light. Sometimes an idea was stated, though not clearly, and then almost forgotten for centuries before it reappeared and was generally accepted.

Pythagoras (~550 BCE) suggested that vision is like touch and that light consists of rays that, acting like feelers, travel in straight lines from the eye to the object and that the sensation of sight is obtained when these rays touch the object.

The Atomists and Plato (5th century BCE) favoured the corpuscular theory of light. Light travels in straight lines and bounces off a mirror like a ball off a wall.

Greeks and Romans reported to use burning glasses 424 BCE.

Aristotle and Euclid (4th century BCE) discussed rays of light.

Euclid (~300 BCE) discussed laws of reflection. Hero of Alexandria thought that light travels by shortest path. Seneca noted that a globe filled with water magnifies.

Epicurus of Samos (~300 BCE) believed that light is emitted by a source and reflected by an object and then enters the eye to produce the sensation of sight. However, the Pythagorean hypothesis generally persisted until Alhazen.

Ptolemy (~50BCE) measured angles of incidence and of refraction. He correctly deduced that the ray is bent toward the normal on entering the denser medium.

EUROPEAN DARK AGES

The Arabic mathematician and physicist Alhazen (1000 CE) published a treatise on optics that was translated into Latin in 1270. In it he developed theories on refraction, reflection, binocular vision, focusing with lenses, the rainbow, parabolic and spherical mirrors, spherical aberration, atmospheric refraction, and the apparent increase in size of planetary bodies near the Earth's horizon. He was first to give an accurate account of vision, correctly stating that light comes from the object seen to the eye.

Roger Bacon (1215-1294) knew of Alhazens work. Studied rays and lenses. Made the earliest recorded comment on lenses for correcting vision (1268), but magnifying lenses inserted in frames were used for reading both in Europe and China at this time.

Early Dutch painters (1300-1400) are believed to have used the camera obsucra and focussing mirrors (according to David Hockney).

THE RENAISSANCE

Leonardo da Vinci (1452-1528) described the "camera obscura".Della Porta (1535 - 1615) camera obscura with a convex lens at the aperture - likened this to the eye.

Kepler (1604) published Dioptrice: discussed rays, optics for thin lens systems and an approximate law of refraction. Light propagated instantaneously. Description of shadows.

Galileo Galilei (1565 - 1642) - first high quality telescope (1609).

Snell (1591 - 1626) empirically rediscovered (1621) and Descartes (1596 - 1650) formulated an exact law of refraction. Descartes: velocity changed when travelling in different media. Light viewed as a pressure transmitted in an elastic medium (1637) i.e. a longitudinal wave.

Pierre Fermat (1601 - 1665) suggested that path taken by a ray of light is "that which takes the least time". An almost correct statement.

Robert Hooke (1635 - 1703) described the microscope 1665.

Ole Chistensen Römer (1644-1710) first measurement of the speed of light (1676).

Christiaan Huygens (1629 - 1695) developed the wave theory and the principle of secondary wavefronts: light moved in an "ether" Traité de la Lumière (1690). Studied polarization and correctly inferred that this was inconsistent with longitudinal waves.

Isaac Newton (1642 - 1726) - dispersion of light through prisms - white light is a mixture of colours. Invented the reflecting telescope. Published Opticks (1704) light was "corpuscular" and had finite velocity. Newton's authority upheld the corpuscular theory against the wave theory long after his death.

NINETEENTH CENTURY

Thomas Young studied interference phenomena (1804). Proved that light was capable of bending around obstacles and was strongly criticized.

Augustin Fresnel (1788 - 1827) considered light as waves and showed that interference effects could be explained by wave theory.

Young, Fresnel, Arago, Fizeau, Lloyd and Kirchhoff all contributed to the establishment of a transverse wave concept for light.

James Clark Maxwell (1831-1879) mathematically unified magnetism and electricity to explain light as an electromagnetic wave.

Michelson and Morley (1887) attempted to detect the motion of the earth with respect to the hypothetical "ether". The null result discredited these theories.

Heinrick Hertz (1857-1894) generated and detected radio waves (1888).

TWENTIETH CENTURY

Max Planck (1858-1947) quantized the radiation field to explain black body radiation (1900).

Albert Einstein (1879-1955) Special relativity (1905): speed of light is a universal constant independent of motion, an "ether" is superfluous.

Einstein - again - proposed particle theory of light to explain the photoelectric effect (1905). Much later he said "I spent my life to find out what a photon is and I still don't know it."

Einstein – once again – wrote down the equations (1916) that formed the theoretical foundation for the laser.

Charles Townes built the first maser (1953). The first laser was demonstrated in 1960.This decade saw the early development of fibre optics and light for communication. Laser light bounced off a mirror left on the moon (1970).Adaptive optics system at the European Southern Observatory produces images to rival the Hubble space telescope (2001).

MODERN PHYSICS

All particles have a dual wave-particle nature. Either view is a simple explanation for something more complex. Which aspect comes to the fore depends on the experiment or phenomenon we are interested in. Particles of light, called photons, are massless and so the classical electromagnetic wave theory is frequently the best to adopt.

Chapter 2

Electromagnetic Waves

At this point in the course we'll move into optics. This might seem like a separate topic from electricity and magnetism, but optics is really a sub-topic of electricity and magnetism. This is because optics deals with the behaviour of light, and light is one example of an electromagnetic wave.

LIGHT AND OTHER ELECTROMAGNETIC WAVES

Light is not the only example of an electromagnetic wave. Other electromagnetic waves include the microwaves you use to heat up leftovers for dinner, and the radio waves that are broadcast from radio stations. An electromagnetic wave can be created by accelerating charges; moving charges back and forth will produce oscillating electric and magnetic fields, and these travel at the speed of light. It would really be more accurate to call the speed "the speed of an electromagnetic wave", because light is just one example of an electromagnetic wave.

speed of light in vacuum: $c = 3.00 \times 10^8$ m/s

As we'll go into later in the course when we get to relativity, c is the ultimate speed limit in the universe. Nothing can travel faster than light in a vacuum.

There is a wonderful connection between c, the speed of light in a vacuum, and the constants that appeared in the electricity and magnetism equations, the permittivity of free space and the permeability of free space. James Clerk Maxwell, who showed that all of electricity and magnetism could be boiled down to four basic equations, also worked out that:

This clearly shows the link between optics, electricity, and magnetism.

CREATING AN ELECTROMAGNETIC WAVE

We've already learned how moving charges (currents) produce magnetic fields. A constant current produces a constant magnetic field, while a changing current produces a changing field. We can go the other way, and use a magnetic field to produce a current, as long as the magnetic field is changing. This is what induced emf is all about. A steadily-changing magnetic field can induce a constant voltage, while an oscillating magnetic field can induce an oscillating voltage.

Focus on these two facts:

1. An oscillating electric field generates an oscillating magnetic field
2. An oscillating magnetic field generates an oscillating electric field

Those two points are key to understanding electromagnetic waves.

An electromagnetic wave (such as a radio wave) propagates outwards from the source (an antenna, perhaps) at the speed of light. What this means in practice is that the source has created oscillating electric and magnetic fields, perpendicular to each other, that travel away from the source. The E and B fields, along with being perpendicular to each other, are perpendicular to the direction the wave travels, meaning that an electromagnetic wave is a transverse wave. The energy of the wave is stored in the electric and magnetic fields.

PROPERTIES OF ELECTROMAGNETIC WAVES

Something interesting about light, and electromagnetic waves in general, is that no medium is required for the wave to travel through. Other waves, such as sound waves, can not travel through a vacuum. An electromagnetic wave is perfectly happy to do that.

An electromagnetic wave, although it carries no mass, does carry energy. It also has momentum, and can exert pressure (known as radiation pressure). The reason tails of comets point away from the Sun is the radiation pressure exerted on the tail by the light (and other forms of radiation) from the Sun.

The energy carried by an electromagnetic wave is proportional to the frequency of the wave. The wavelength and frequency of the wave are connected via the speed of light:

Electromagnetic waves are split into different categories based on their frequency (or, equivalently, on their wavelength). In other words, we split up the electromagnetic spectrum based on frequency. Visible light, for example, ranges from violet to red. Violet light has a wavelength of 400 nm, and a frequency of 7.5×10^{14} Hz. Red light has a wavelength of 700 nm, and a frequency of 4.3×10^{14} Hz. Any electromagnetic wave with a frequency (or wavelength) between those extremes can be seen by humans.

Visible light makes up a very small part of the full electromagnetic spectrum. Electromagnetic waves that are of higher energy than visible light (higher frequency, shorter wavelength) include ultraviolet light, X-rays, and gamma rays. Lower energy waves (lower frequency, longer wavelength) include infrared light, microwaves, and radio and television waves.

ENERGY IN AN ELECTROMAGNETIC WAVE

The energy in an electromagnetic wave is tied up in the electric and magnetic fields. In general, the energy per unit volume in an electric field is given by:

In a magnetic field, the energy per unit volume is:

An electromagnetic wave has both electric and magnetic fields, so the total energy density associated with an electromagnetic wave is:

It turns out that for an electromagnetic wave, the energy associated with the electric field is equal to the energy associated with the magnetic field, so the energy density can be written in terms of just one or the other:

This also implies that in an electromagnetic wave, E = cB.

A more common way to handle the energy is to look at how much energy is carried by the wave from one place to another. A good measure of this is the intensity of the wave, which is the power that passes perpendicularly through an area divided by the area. The intensity, S, and the energy density are related by a factor of c:

Generally, it's most useful to use the average power, or average intensity, of the wave. To find the average values, you have to use some average for the electric field E and the magnetic field B. The root mean square averages are used; the relationship between the peak and rms values is:

THE REFLECTION AND REFRACTION OF LIGHT

RAYS AND WAVE FRONTS

Light is a very complex phenomenon, but in many situations its behaviour can be understood with a simple model based on rays and wave fronts. A ray is a thin beam of light that travels in a straight line. A wave front is the line (not necessarily straight) or surface connecting all the light that left a source at the same time. For a source like the Sun, rays radiate out in all directions; the wave fronts are spheres centred on the Sun. If the source is a long way away, the wave fronts can be treated as parallel lines.

Rays and wave fronts can generally be used to represent light when the light is interacting with objects that are much larger than the wavelength of light, which is about 500 nm. In particular, we'll use rays and wave fronts to analyse how light interacts with mirrors and lenses.

THE LAW OF REFLECTION

Objects can be seen by the light they emit, or, more often, by the light they reflect. Reflected light obeys the law of

reflection, that the angle of reflection equals the angle of incidence.

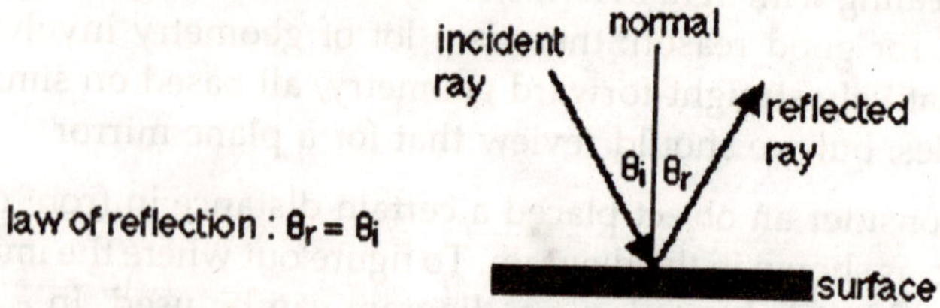

For objects such as mirrors, with surfaces so smooth that any hills or valleys on the surface are smaller than the wavelength of light, the law of reflection applies on a large scale. All the light travelling in one direction and reflecting from the mirror is reflected in one direction; reflection from such objects is known as specular reflection.

Most objects exhibit diffuse reflection, with light being reflected in all directions. All objects obey the law of reflection on a microscopic level, but if the irregularities on the surface of an object are larger than the wavelength of light, which is usually the case, the light reflects off in all directions.

PLANE MIRRORS

A plane mirror is simply a mirror with a flat surface; all of us use plane mirrors every day, so we've got plenty of experience with them. Images produced by plane mirrors have a number of properties, including:

1. The image produced is upright
2. The image is the same size as the object (i.e., the magnification is m = 1)
3. The image is the same distance from the mirror as the object appears to be (i.e., the image distance = the object distance)
4. The image is a virtual image, as opposed to a real image, because the light rays do not actually pass through the image. This also implies that an image could not be focused on a screen placed at the location where the image is.

A little geometry

Dealing with light in terms of rays is known as geometrical optics, for good reason: there is a lot of geometry involved. It's relatively straight-forward geometry, all based on similar triangles, but we should review that for a plane mirror.

Consider an object placed a certain distance in front of a mirror, as shown in the diagram. To figure out where the image of this object is located, a ray diagram can be used. In a ray diagram, rays of light are drawn from the object to the mirror, along with the rays that reflect off the mirror. The image will be found where the reflected rays intersect. Note that the reflected rays obey the law of reflection. What you notice is that the reflected rays diverge from the mirror; they must be extended back to find the place where they intersect, and that's where the image is.

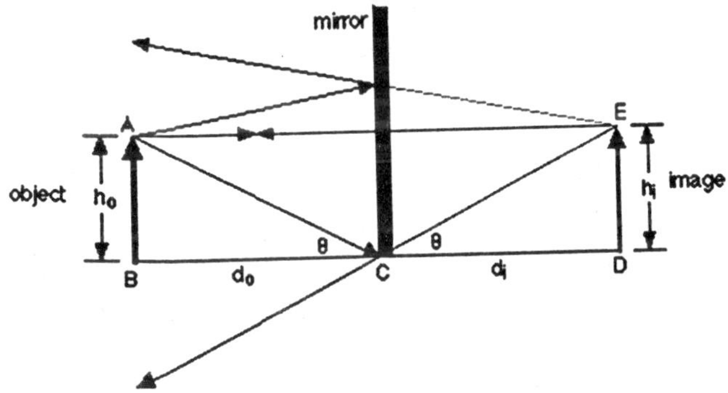

Analyzing this a little further, it's easy to see that the height of the image is the same as the height of the object. Using the similar triangles ABC and EDC, it can also be seen that the distance from the object to the mirror is the same as the distance from the image to the mirror.

SPHERICAL MIRRORS

Light reflecting off a flat mirror is one thing, but what happens when light reflects off a curved surface? We'll take a look at what happens when light reflects from a spherical mirror, because it turns out that, using reasonable

approximations, this analysis is fairly straight-forward. The image you see is located either where the reflected light converges, or where the reflected light appears to diverge from.

A spherical mirror is simply a piece cut out of a reflective sphere. It has a centre of curvature, C, which corresponds to the centre of the sphere it was cut from; a radius of curvature, R, which corresponds to the radius of the sphere; and a focal point (the point where parallel light rays are focused to) which is located half the distance from the mirror to the centre of curvature. The focal length, f, is therefore:

focal length of a spherical mirror: $f = R / 2$

This is actually an approximation. Parabolic mirrors are really the only mirrors that focus parallel rays to a single focal point, but as long as the rays don't get too far from the principal axis then the equation above applies for spherical mirrors. The diagram shows the principal axis, focal point (F), and centre of curvature for both a concave and convex spherical mirror.

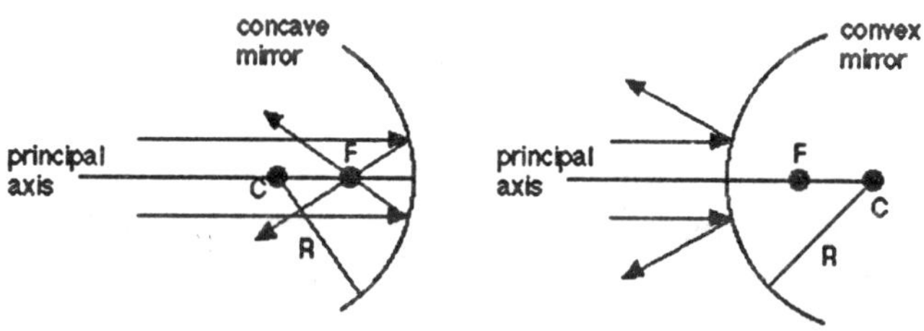

Spherical mirrors are either concave (converging) mirrors or convex (diverging) mirrors, depending on which side of the spherical surface is reflective. If the inside surface is reflective, the mirror is concave; if the outside is reflective, it's a convex mirror. Concave mirrors can form either real or virtual images, depending on where the object is relative to the focal point. A convex mirror can only form virtual images. A real image is an image that the light rays from the object actually pass through; a virtual image is formed because the light rays can be extended back to meet at the image position, but they don't actually go through the image position.

RAY DIAGRAMS

To determine where the image is, it is very helpful to draw a ray diagram. The image will be located at the place where the rays intersect. You could just draw random rays from the object to the mirror and follow the reflected rays, but there are three rays in particular that are very easy to draw.

Only two rays are necessary to locate the image on a ray diagram, but it's useful to add the third as a check. The first is the parallel ray; it is drawn from the tip of the object parallel to the principal axis. It then reflects off the mirror and either passes through the focal point, or can be extended back to pass through the focal point.

The second ray is the chief ray. This is drawn from the tip of the object to the mirror through the centre of curvature. This ray will hit the mirror at a 90° angle, reflecting back the way it came. The chief and parallel rays meet at the tip of the image.

The third ray, the focal ray, is a mirror image of the parallel ray. The focal ray is drawn from the tip of the object through (or towards) the focal point, reflecting off the mirror parallel to the principal axis. All three rays should meet at the same point.

A ray diagram for a concave mirror is shown above. This shows a few different things. For this object, located beyond the centre of curvature from the mirror, the image lies between the focal point (F) and the centre of curvature. The image is inverted compared to the object, and it is also a real image, because the light rays actually pass through the point where the image is located.

With a concave mirror, any object beyond C will always have an image that is real, inverted compared to the object, and between F and C. You can always trade the object and image places (that just reverses all the arrows on the ray diagram), so any object placed between F and C will have an image that is real, inverted, and beyond C. What happens when the object is between F and the mirror? You should draw the ray diagram to convince yourself that the image will be behind the mirror, making it a virtual image, and it will be upright compared to the object.

A RAY DIAGRAM FOR A CONVEX MIRROR

What happens with a convex mirror? In this case the ray diagram looks like this:

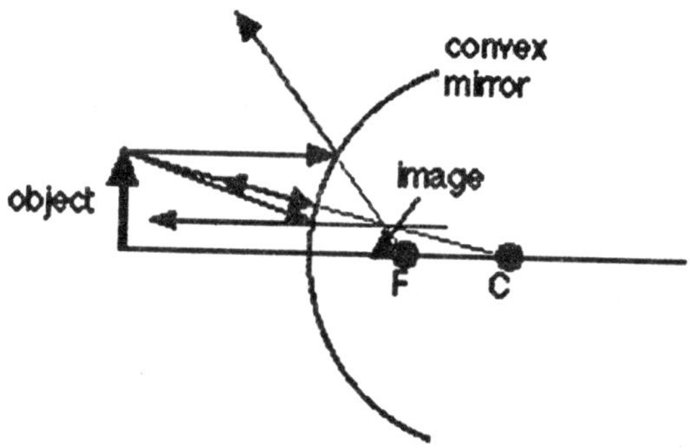

As the ray diagram shows, the image for a convex mirror is virtual, and upright compared to the object. A convex mirror is the kind of mirror used for security in stores, and is also the kind of mirror used on the passenger side of many cars ("Objects in mirror are closer than they appear."). A convex mirror will reflect a set of parallel rays in all directions; conversely, it will also take light from all directions and reflect it in one direction, which is exactly how it's used in stores and cars.

The Mirror Equation

Drawing a ray diagram is a great way to get a rough idea of how big the image of an object is, and where the image is located. We can also calculate these things precisely, using something known as the mirror equation. The textbook does a nice job of deriving this equation in section 25.6, using the geometry of similar triangles.

Magnification

In most cases the height of the image differs from the height of the object, meaning that the mirror has done some magnifying (or reducing). The magnification, m, is defined as the ratio of the image height to the object height, which is closely related to the ratio of the image distance to the object distance:

A magnification of 1 (plus or minus) means that the image is the same size as the object. If m has a magnitude greater than 1 the image is larger than the object, and an m with a magnitude less than 1 means the image is smaller than the object. If the magnification is positive, the image is upright compared to the object; if m is negative, the image is inverted compared to the object.

Sign Conventions

What does a positive or negative image height or image distance mean? To figure out what the signs mean, take the side of the mirror where the object is to be the positive side. Any distances measured on that side are positive. Distances measured on the other side are negative.

F, the focal length, is positive for a concave mirror, and negative for a convex mirror.

When the image distance is positive, the image is on the same side of the mirror as the object, and it is real and inverted. When the image distance is negative, the image is behind the mirror, so the image is virtual and upright.

A negative m means that the image is inverted. Positive means an upright image.

STEPS FOR ANALYZING MIRROR PROBLEMS

There are basically three steps to follow to analyse any mirror problem, which generally means determining where the image of an object is located, and determining what kind of image it is (real or virtual, upright or inverted).

- *Step 1* - Draw a ray diagram. The more careful you are in constructing this, the better idea you'll have of where the image is.
- *Step 2* - Apply the mirror equation to determine the image distance. (Or to find the object distance, or the focal length, depending on what is given.)

- *Step 3* - Make sure steps 1 and 2 are consistent with each other.

REFRACTION

When we talk about the speed of light, we're usually talking about the speed of light in a vacuum, which is 3.00 x 10^8 m/s. When light travels through something else, such as glass, diamond, or plastic, it travels at a different speed. The speed of light in a given material is related to a quantity called the index of refraction, n, which is defined as the ratio of the speed of light in vacuum to the speed of light in the medium:

Index of refraction: $n = c / v$

When light travels from one medium to another, the speed changes, as does the wavelength. The index of refraction can also be stated in terms of wavelength:

Although the speed changes and wavelength changes, the frequency of the light will be constant. The frequency, wavelength, and speed are related by:

The change in speed that occurs when light passes from one medium to another is responsible for the bending of light, or refraction, that takes place at an interface. If light is travelling from medium 1 into medium 2, and angles are measured from the normal to the interface, the angle of transmission of the light into the second medium is related to the angle of incidence by Snell's law:

TOTAL INTERNAL REFLECTION, AND LENSES

TOTAL INTERNAL REFLECTION

When light crosses an interface into a medium with a higher index of refraction, the light bends towards the normal. Conversely, light traveling across an interface from higher n to lower n will bend away from the normal. This has an interesting implication: at some angle, known as the critical angle, light travelling from a medium with higher n to a medium with lower n will be refracted at 90°; in other words, refracted along the interface. If the light hits the interface at any angle larger than this critical angle, it will not pass through to the second medium at all. Instead, all of it will be reflected back into the first medium, a process known as total internal reflection.

The critical angle can be found from Snell's law, putting in an angle of 90° for the angle of the refracted ray. This gives:

For any angle of incidence larger than the critical angle, Snell's law will not be able to be solved for the angle of refraction, because it will show that the refracted angle has a sine larger than 1, which is not possible. In that case all the light is totally reflected off the interface, obeying the law of reflection.

Optical fibers are based entirely on this principle of total internal reflection. An optical fiber is a flexible strand of glass. A fiber optic cable is usually made up of many of these strands, each carrying a signal made up of pulses of laser light. The light travels along the optical fiber, reflecting off the walls of the fiber. With a straight or smoothly bending fiber, the light will hit the wall at an angle higher than the critical angle and will all be reflected back into the fiber. Even though the light undergoes a large number of reflections when traveling along a fiber, no light is lost.

DEPTH PERCEPTION

Because light is refracted at interfaces, objects you see across an interface appear to be shifted relative to where they really are. If you look straight down at an object at the bottom of a glass of water, for example, it looks closer to you than it really is. Looking perpendicular to an interface, the apparent depth is related to the actual depth by:

apparent depth: $d' = d(n_2/n_1)$

LENSES

There are many similarities between lenses and mirrors. The mirror equation, relating focal length and the image and object distances for mirrors, is the same as the lens equation used for lenses.There are also some differences, however; the most important being that with a mirror, light is reflected, while with a lens an image is formed by light that is refracted by, and transmitted through, the lens. Also, lenses have two focal points, one on each side of the lens.

The surfaces of lenses, like spherical mirrors, can be treated as pieces cut from spheres. A lens is double sided, however, and the two sides may or may not have the same curvature. A general rule of thumb is that when the lens is thickest in the centre, it is a converging lens, and can form real or virtual images. When the lens is thickest at the outside, it is a diverging lens, and it can only form virtual images.

Consider first a converging lens, the most common type being a double convex lens. As with mirrors, a ray diagram should be drawn to get an idea of where the image is and what the image characteristics are. Drawing a ray diagram for a lens is very similar to drawing one for a mirror. The parallel ray is drawn the same way, parallel to the optic axis, and through (or extended back through) the focal point. The chief difference in the ray diagram is with the chief ray. That is drawn from the tip of the object straight through the centre of the lens. Wherever the two rays meet is where the image is. The third ray, which can be used as a check, is drawn from the tip of the object through the focal point that is on the same side of the lens as the object. That ray travels through the lens, and is refracted so it travels parallel to the optic axis on the other side of the lens.

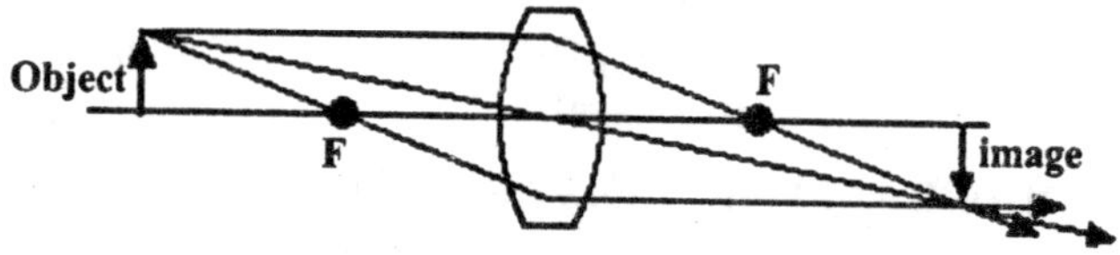

Fig. The rays actually bend twice, once at each air-lens interface, but for a thin lens we can approximate what the light does with a single refreaction for the parallel and focal rays, and no reftraction at all for the chief ray.

The two sides of the lens are referred to as the object side (the side where the object is) and the image side. For lenses, a positive image distance means that the image is real and is on the image side. A negative image distance means the image is on the same side of the lens as the object; this must be a virtual image.

lens equation: $1/d_i = 1/f$

Using that sign convention gives a lens equation identical to the spherical mirror equation:

The other signs work the same way as for mirrors. The focal length, f, is positive for a converging lens, and negative for a diverging lens.

The magnification factor is also given by the same equation:

magnification: $m = h_i/h_0 = -d_i/d_0$

Useful devices like microscopes and telescopes rely on at least two lenses (or mirrors). A microscope, for example, is a compound lens system with two converging lenses. One important thing to note is that with two lenses (and you can extend the argument for more than two), the magnification factor m for the two lens system is the product of the two individual magnification factors.

It works this way. The first lens takes an object and creates an image at a particular point, with a certain magnification factor (say, 3 times as large). The second lens uses the image created by the first lens as the object, and creates a final image, introducing a second magnification factor (say, a factor of seven). The magnification factor of the final image compared to the original object is the product of the two magnification factors (3 × 7 = 21, in this case).

RAY DIAGRAM FOR A DIVERGING LENS

Consider now the ray diagram for a diverging lens. Diverging lenses come in a few different shapes, but all diverging lens are fatter on the edge than they are in the centre. A good example of a diverging lens is a bi-concave lens, as shown in the diagram. The object in this case is beyond the focal point, and, as usual, the place where the refracted rays appear to diverge from is the location of the image. A diverging lens always gives a virtual image, because the refracted rays have to be extended back to meet.

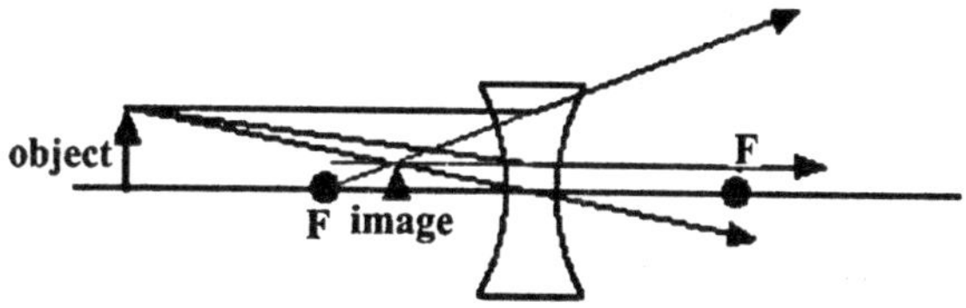

Note that a diverging lens will refract parallel rays so that they diverge from each other, while a converging lens refracts parallel rays toward each other.

INTERFERENCE

THE WAVE NATURE OF LIGHT

When we discussed the reflection and refraction of light, light was interacting with mirrors and lenses. These objects are much larger than the wavelength of light, so the analysis can be done using geometrical optics, a simple model that uses rays and wave fronts. In this chapter we'll need a more sophisticated model, physical optics, which treats light as a wave. The wave properties of light are important in understanding how light interacts with objects such as narrow openings or thin films that are about the size of the wavelength of light.

Because physical optics deals with light as a wave, it is helpful to have a quick review of waves. The principle of linear superposition is particularly important.

LINEAR SUPERPOSITION

When two or more waves come together, they will interfere with each other. This interference may be constructive or destructive. If you take two waves and bring them together, they will add wherever a peak from one matches a peak from the other. That's constructive interference. Wherever a peak from one wave matches a trough in another wave, however, they will cancel each other out (or partially cancel, if the amplitudes are different); that's destructive interference.

The most interesting cases of interference usually involve identical waves, with the same amplitude and wavelength, coming together. Consider the case of just two waves, although we can generalize to more than two. If these two waves come from the same source, or from sources that are emitting waves in phase, then the waves will interfere constructively at a certain point if the distance traveled by one wave is the same as, or differs by an integral number of wavelengths from, the path length traveled by the second wave. For the waves to interfere destructively, the path lengths must differ by an integral number of wavelengths plus half a wavelength.

condition for constructive interference: path length difference = λ (m=0, 1, 2, 3,...)

condition for constructive interference: path length difference = $(m + 1/2)\lambda$

YOUNG'S DOUBLE SLIT

Light, because of its wave properties, will show constructive and destructive interference. This was first shown in 1801 by Thomas Young, who sent sunlight through two narrow slits and showed that an interference pattern could be seen on a screen placed behind the two slits. The interference pattern was a set of alternating bright and dark lines, corresponding to where the light from one slit was alternately constructively and destructively interfering with the light from the second slit.

You might think it would be easier to simply set up two light sources and look at their interference pattern, but the phase relationship between the waves is critically important, and two sources tend to have a randomly varying phase relationship. With a single source shining on two slits, the relative phase of the light emitted from the two slits is kept constant.

This makes use of Huygen's principle, the idea that each point on a wave can be considered to be a source of secondary waves. Applying this to the two slits, each slit acts as a source of light of the same wavelength, with the light from the two slits interfering constructively or destructively to produce an interference pattern of bright and dark lines.

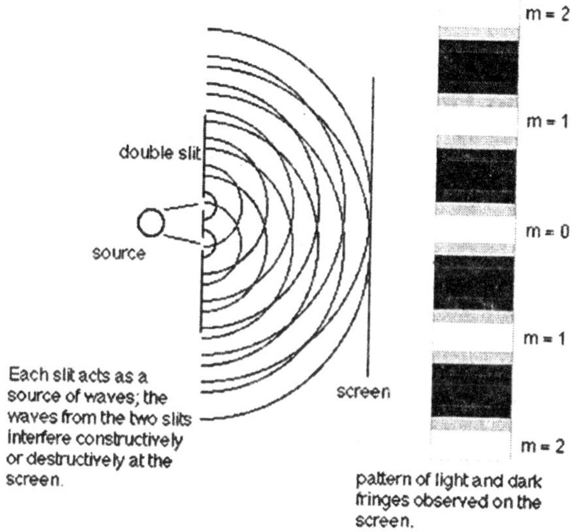

This pattern of bright and dark lines is known as a fringe pattern, and is easy to see on a screen. The bright fringe in the middle is caused by light from the two slits traveling the same distance to the screen; this is known as the zero-order fringe.The dark fringes on either side of the zero-order fringe are caused by light from one slit traveling half a wavelength further than light from the other slit. These are followed by the first-order fringes (one on each side of the zero-order fringe), caused by light from one slit traveling a wavelength further than light from the other slit, and so on.

The diagram above shows the geometry for the fringe pattern. For two slits separated by a distance d, and emitting light at a particular wavelength, light will constructively interfere at certain angles. These angles are found by applying the condition for constructive interference, which in this case becomes:

bright fringes of a double slit: $d\sin\theta = m\lambda$ ($m = 0, 1, 2, 3,...$)

The angles at which dark fringes occur can be found be applying the condition for destructive interference:

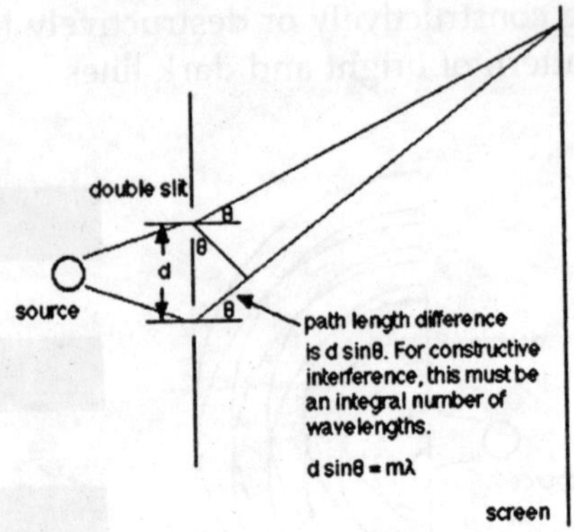

dark fringes of a double slit: dsinθ = (m + 1/2)λ (m = 0, 1, 2, 3,...)

If the interference pattern was being viewed on a screen a distance L from the slits, the wavelength can be found from the equation:

$$\lambda = yd/(mL)$$

where y is the distance from the centre of the interference pattern to the mth bright line in the pattern. That applies as long as the angle is small (i.e., y must be small compared to L).

DISPERSION

Although we talk about an index of refraction for a particular material, that is really an average value. The index of refraction actually depends on the frequency of light (or, equivalently, the wavelength). For visible light, light of different colours means light of different wavelength. Red light has a wavelength of about 700 nm, while violet, at the other end of the visible spectrum, has a wavelength of about 400 nm.

This doesn't mean that all violet light is at 400 nm. There are different shades of violet, so violet light actually covers a range of wavelengths near 400 nm. Likewise, all the different shades of red light cover a range near 700 nm.

Because the refractive index depends on the wavelength, light of different colours (i.e., wavelengths) travels at different speeds in a particular material, so they will be refracted through slightly different angles inside the material. This is called dispersion, because light is dispersed into colours by the material.

When you see a rainbow in the sky, you're seeing something produced by dispersion and internal reflection of light in water droplets in the atmosphere. Light from the sun enters a spherical raindrop, and the different colours are refracted at different angles, reflected off the back of the drop, and then bent again when they emerge from the drop. The different colours, which were all combined in white light, are now dispersed and travel in slightly different directions. You see red light coming from water droplets higher in the sky than violet light. The other colours are found between these, making a rainbow.

Rainbows are usually seen as half circles. If you were in a plane or on a very tall building or mountain, however, you could see a complete circle. In double rainbows the second, dimmer, band, which is higher in the sky than the first, comes from light reflected twice inside a raindrop. This reverses the order of the colours in the second band.

DIFFRACTION; THIN-FILM INTERFERENCE

DIFFRACTION

We discussed diffraction in PY105 when we talked about sound waves; diffraction is the bending of waves that occurs when a wave passes through a single narrow opening. The analysis of the resulting diffraction pattern from a single slit is similar to what we did for the double slit. With the double slit, each slit acted as an emitter of waves, and these waves interfered with each other. For the single slit, each part of the slit can be thought of as an emitter of waves, and all these waves interfere to produce the interference pattern we call the diffraction pattern.

After we do the analysis, we'll find that the equation that gives the angles at which fringes appear for a single slit is very similar to the one for the double slit, one obvious difference being that the slit width (W) is used in place of d, the distance between slits. A big difference between the single and double slits, however, is that the equation that gives the bright fringes for the double slit gives dark fringes for the single slit.

To see why this is, consider the diagram below, showing light going away from the slit in one particular direction.

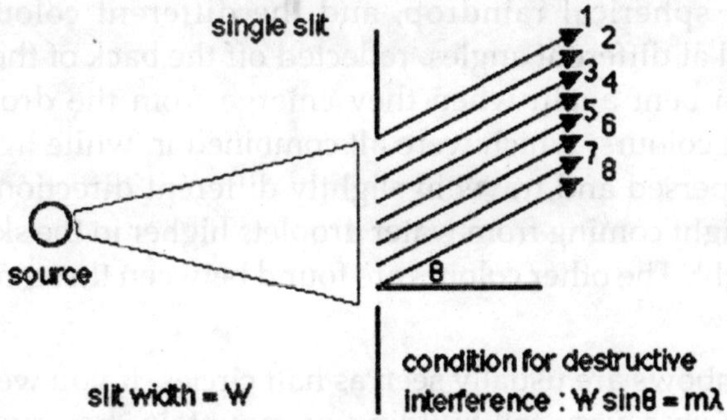

In the diagram above, let's say that the light leaving the edge of the slit (ray 1) arrives at the screen half a wavelength out of phase with the light leaving the middle of the slit (ray 5). These two rays would interfere destructively, as would rays 2 and 6, 3 and 7, and 4 and 8. In other words, the light from one half of the opening cancels out the light from the other half. The rays are half a wavelength out of phase because of the extra path length traveled by one ray; in this case that extra distance is:

$(w/2)\sin\theta = \lambda/2$

The factors of 2 cancel, leaving:

$W \sin\theta = \lambda$

The argument can be extended to show that:

dark fringes of a single slit: $W\sin\theta = m\lambda$ $(m = 0, 1, 2, 3,...)$

The bright fringes fall between the dark ones, with the central bright fringe being twice as wide, and considerably brighter, than the rest.

DIFFRACTION EFFECTS WITH A DOUBLE SLIT

Note that diffraction can be observed in a double-slit interference pattern. Essentially, this is because each slit emits a diffraction pattern, and the diffraction patterns interfere with each other. The shape of the diffraction pattern is determined by the width (W) of the slits, while the shape of the interference pattern is determined by d, the distance between the slits. If W is much larger than d, the pattern will be dominated by interference effects; if W and d are about the same size the two effects will contribute equally to the fringe pattern. Generally what you see is a fringe pattern that has missing interference fringes; these fall at places where dark fringes occur in the diffraction pattern.

DIFFRACTION GRATINGS

We've talked about what happens when light encounters a single slit (diffraction) and what happens when light hits a double slit (interference); what happens when light encounters an entire array of identical, equally-spaced slits? Such an array is known as a diffraction grating. The name is a bit misleading, because the structure in the pattern observed is dominated by interference effects.

With a double slit, the interference pattern is made up of wide peaks where constructive interference takes place. As more slits are added, the peaks in the pattern become sharper and narrower. With a large number of slits, the peaks are very sharp. The positions of the peaks, which come from the constructive interference between light coming from each slit, are found at the same angles as the peaks for the double slit; only the sharpness is affected.

Angles at which constructive interference occurs: $d\sin\theta = m\lambda$ ($m = 0, 1, 2, 3,...$)

Why is the pattern much sharper? In the double slit, between each peak of constructive interference is a single location where destructive interference takes place. Between the central peak ($m = 0$) and the next one ($m = 1$), there is a place where one wave travels 1/2 a wavelength further than the other, and that's where destructive interference takes place. For three slits, however, there are two places where destructive interference takes place. One is located at the point where the path lengths differ by 1/3 of a wavelength, while the other is at the place where the path lengths differ by 2/3 of a wavelength. For 4 slits, there are three places, for 5 slits there are four places, etc. Completely constructive interference, however, takes place only when the path lengths differ by an integral number of wavelengths. For a diffraction grating, then, with a large number of slits, the pattern is sharp because of all the destructive interference taking place between the bright peaks where constructive interference takes place.

Diffraction gratings, like prisms, disperse white light into individual colours. If the grating spacing (d, the distance between slits) is known and careful measurements are made of the angles at which light of a particular colour occurs in the interference pattern, the wavelength of the light can be calculated.

THIN-FILM INTERFERENCE

Interference between light waves is the reason that thin films, such as soap bubbles, show colourful patterns. This is known as thin-film interference, because it is the interference of light waves reflecting off the top surface of a film with the waves reflecting from the bottom surface. To obtain a nice coloured pattern, the thickness of the film has to be similar to the wavelength of light.

An important consideration in determining whether these waves interfere constructively or destructively is the fact that whenever light reflects off a surface of higher index of refraction, the wave is inverted. Peaks become troughs, and troughs become peaks. This is referred to as a 180° phase shift in the wave, but the easiest way to think of it is as an effective shift in the wave by half a wavelength.

Summarizing this, reflected waves experience a 180° phase shift (half a wavelength) when reflecting from a higher-n medium ($n_2 > n_1$), and no phase shift when reflecting from a medium of lower index of refraction ($n_2 < n_1$).

For completely constructive interference to occur, the two reflected waves must be shifted by an integer multiple of wavelengths relative to one another. This relative shift includes any phase shifts introduced by reflections off a higher-n medium, as well as the extra distance traveled by the wave that goes down and back through the film.

Note that one has to be very careful in dealing with the wavelength, because the wavelength depends on the index of refraction. Generally, in dealing with thin-film interference the key wavelength is the wavelength in the film itself. If the film has an index of refraction n, this wavelength is related to the wavelength in vacuum by:

$$\lambda_{film} = \lambda_{vacuum/n}$$

A STEP-BY STEP APPROACH

Many people have trouble with thin-film interference problems. As usual, applying a systematic, step-by-step approach is best. The overall goal is to figure out the shift of the wave reflecting from one surface of the film relative to the wave that reflects off the other surface. Depending on the situation, this shift is set equal to the condition for constructive interference, or the condition for destructive interference.

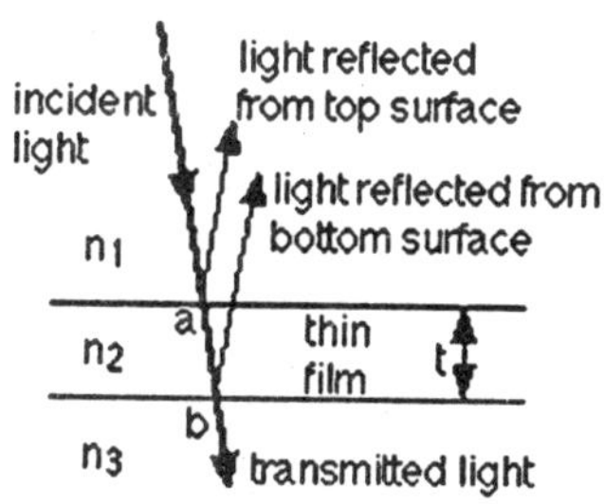

Fig. The diagram shows a typical arrangement for thin-film interference. Light travelling in one medium encounters a thin film with index of refraction n_2. Some of the light is transmitted through the film. Thin-film interference involves interference between ligh reflecting off the top surface of the film (at a) and light reflecting from the bottom surface of the film(at b).

Note that typical thin-film interference problems involve "normally-incident" light. The light rays are not drawn perpendicular to the interfaces on the diagram to make it easy to distinguish between the incident and reflected rays. In the discussion below it is assumed that the incident and reflected rays are perpendicular to the interfaces.

A good method for analyzing a thin-film problem involves these steps:

Step 1. Write down Δ_a, the shift for the wave reflecting off the top surface of the film.

This shift will either be 0, if $n_1 > n_2$, if $n_1 < n_2$. So, $\Delta_a = 0$ or $\Delta_a = \lambda/2$.

Step 2. Write down, Δ_b the shift for the wave reflecting off the film's bottom surface.

One contribution to this shift comes from the extra distance travelled. If the film thickness is t, this wave goes down and back through the film, so its path length is longer by 2t. The other contribution to this shift can be either 0 or, depending on what happens when it reflects (this reflection occurs at point b on the diagram).

$\Delta_b = 2t$ if $n_2 > n_3$, or $\Delta_b = 2t + \lambda/2$ if $n_2 < n_3$.

Step 3. Calculate the relative shift by subtracting the individual shifts.

$$\Delta = \Delta_b - \Delta_a$$

Step 4. Set the relative shift equal to the condition for constructive interference, or the condition for destructive interference, depending on the situation. If a certain film looks red in reflected light, for instance, that means we have constructive interference for red light. If the film is dark, the light must be interfering destructively.

For constructive interference: $\Delta = m\lambda$ (m = an integer)

For destructive interference: $\Delta = (m + 1/2)\lambda$

Step 5. Rearrange the equation (if necessary) to get all factors of on one side.

Step 6. Remember that the wavelength in your equation is the wavelength in the film itself. Since the film is medium 2 in the diagram above, we can label it. The wavelength in the film is related to the wavelength in vacuum by:

$$\lambda_2 = \lambda_{vac/n2}$$

Step 7. Solve. Your equation should give you a relationship between t, the film thickness, and either the wavelength in vacuum or the wavelength in the film.

A FILM OF OIL ON WATER

Working through an example is a good way to see how the step-by-step approach is applied. In this case, white light in air shines on an oil film that floats on water. When looking straight down at the film, the reflected light is red, with a wavelength of 636 nm. What is the minimum possible thickness of the film?

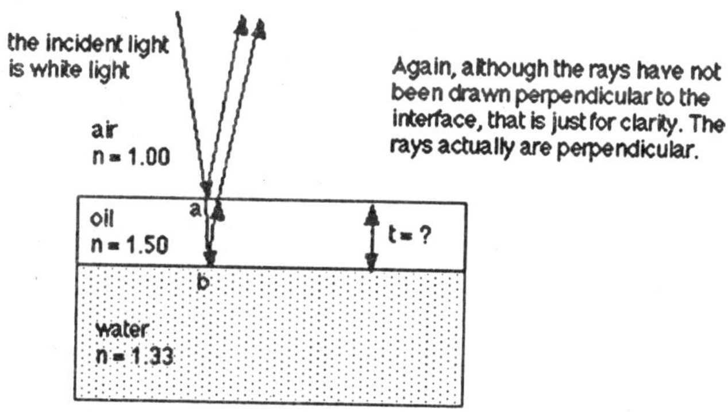

Step 1. Because oil has a higher index of refraction than air, the wave reflecting off the top surface of the film is shifted by half a wavelength.

$$\Delta_a = \lambda/2$$

Step 2. Because water has a lower index of refraction than oil, the wave reflecting off the bottom surface of the film does not have a half-wavelength shift, but it does travel the extra distance of 2t.

$$\Delta_b = 2t$$

Step 3. The relative shift is thus:

$\Delta = \Delta_b - \Delta_a = 2t - \lambda/2$

Step 4. Now, is this constructive interference or destructive interference? Because the film looks red, there is constructive interference taking place for the red light.

For constructive interference: $\Delta = m\lambda$. In this case, $2t - \lambda/2 = m\lambda$

Step 5. Moving all factors of the wavelength to the right side of the equation gives:

$2t = (m + 1/2)\lambda$ (m = 0, 1, 2, 3,...)

Note that this looks like an equation for destructive interference! It isn't, because we used the condition for constructive interference in step 4. It looks like a destructive interference equation only because one reflected wave experienced a $\lambda/2$ shift.

Step 6. The wavelength in the equation above is the wavelength in the thin film. Writing the equation so this is obvious can be done in a couple of different ways:

$2t = (m + 1/2)\lambda 2$ or $2t = (m + 1/2)\lambda_{vac} / n_2$

Step 7. The equation can now be solved. In this situation, we are asked to find the minimum thickness of the film. This means choosing the minimum value of m, which in this case is m = 0. The question specified the wavelength of red light in vacuum, so:

$2t_{min} = (1/2)\ \lambda_{vac} / n_2$

This gives $t_{min} = \lambda_{vac} / 4n_2 = 636 / (4 \times 1.5) = 636 / 6 = 106$ nm

This is not the only thickness that gives completely constructive interference for this wavelength. Others can be found by using m = 1, m = 2, etc. in the equation in step 6.

If 106 nm gives constructive interference for red light, what about the other colours? They are not completely cancelled out, because 106 nm is not the right thickness to give completely destructive interference for any wavelength in the visible spectrum. The other colours do not reflect as intensely as red light, so the film looks red.

Why is it the wavelength in the film itself that matters?

The light reflecting off the top surface of the film does not pass through the film at all, so how can it be the wavelength in the film that is important in thin-film interference? A diagram can help clarify this. The diagram looks a little complicated at first glance, but it really is straightforward once you understand what it shows.

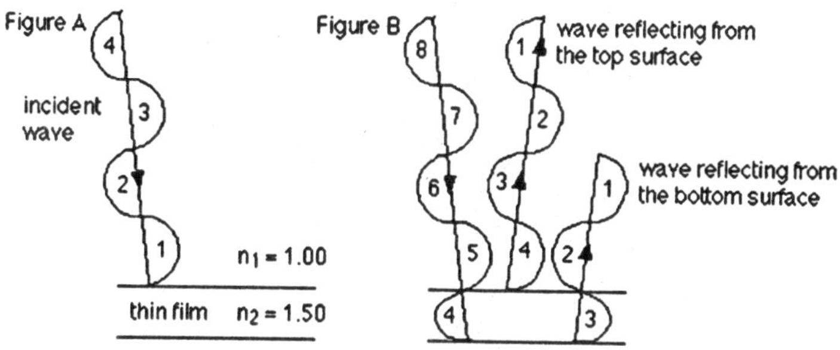

The film thickness is $0.5\ \lambda_2$ $n_3 = 1.33$

Figure B shows the situation two periods after Figure A. Note that both reflected waves have been moved to the right for clarity.

In this situation, the thickness of the thin film is exactly half the wavelength of the wave when it is in the film. This will result in completely destructive interference, as the two reflected waves cancel each other out.

Figure A shows a wave incident on a thin film. Each half wavelength has been numbered, so we can keep track of it. Note that the thickness of the film is exactly half the wavelength of the wave when it is in the film.

Figure B shows the situation two periods later, after two complete wavelengths have encountered the film. Part of the wave is reflected off the top surface of the film; note that this reflected wave is flipped by 180°, so peaks are now troughs and troughs are now peaks. This is because the wave is reflecting off a higher-n medium.

Another part of the wave reflects off the bottom surface of the film. This does not flip the wave, because the reflection is from a lower-n medium. When this wave re-emerges into the first medium, it destructively interferes with the wave that reflects off the top surface. This occurs because the film thickness is exactly half the wavelength of the wave in the film. Because a half wavelength fits in the film, the peaks of one reflected wave line up precisely with the troughs of the other (and vice versa), so the waves cancel. Destructive interference would also occur with the film thickness being equal to 1 wavelength of the wave in the film, or 1.5 wavelengths, 2 wavelengths, etc.

If the thickness was 1/4, 3/4, 5/4, etc. the wavelength in the film, constructive interference occurs. This is only true when one of the reflected waves experiences a half wavelength shift (because of the relative sizes of the refractive indices). If neither wave, or both waves, experiences a shift of, there would be constructive interference whenever the film thickness was 0.5, 1, 1.5, 2, etc. wavelengths, and destructive interference if the film was 1/4, 3/4, 5/4, etc. of the wavelength in the film.

One final philosophical note, to really make your head spin if it isn't already. In the diagram above we drew the two reflected waves and saw how they cancelled out. This means none of the wave energy is reflected back into the first medium. Where does it go? It must all be transmitted into the third medium (that's the whole point of a non-reflective coating, to transmit as much light as possible through a lens). So, even though we did the analysis by drawing the waves reflecting back, in some sense they really don't reflect back at all, because all the light ends up in medium 3.

NON-REFLECTIVE COATINGS

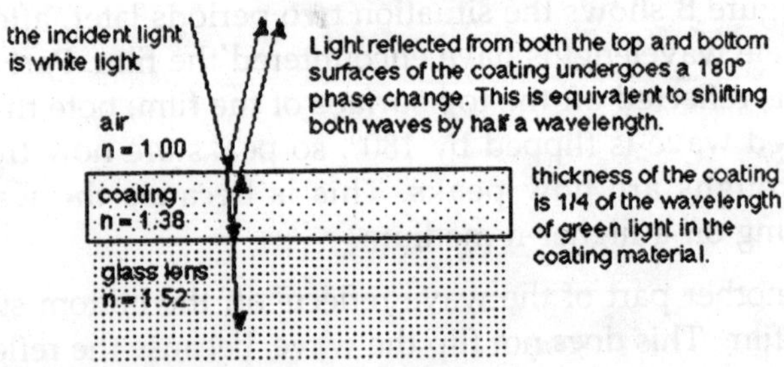

Light reflecting from the bottom surface travels an extra distance of half the wavelength of green light. Green light reflected from the top surface interferes destructively with green light from the bottom surface, in other words, all the green light, and most of the light in the middle of the visible spectrum, is transmitted. Some red and violet light is reflefted, so the coating looks purple.

Destructive interference is exploited in making non-reflective coatings for lenses. The coating material generally has an index of refraction less than that of glass, so both reflected waves have a shift. A film thickness of 1/4 the wavelength in the film results in destructive interference (this is derived below)

For non-reflective coatings in a case like this, where the index of refraction of the coating is between the other two indices of refraction, the minimum film thickness can be found by applying the step-by-step approach:

$\Delta_a = \lambda/2$ $\qquad$ $\Delta_b = 2t + \lambda/2$ $\qquad$ $\Delta = 2t = (m + 1/2)\,\lambda_{coating}$

so

$t_{min} = \lambda_{coating}/4 = \lambda_{vacuum}/(4\,n)$ $\qquad$ n is the refractive index of the coating

POLARIZATION; AND THE HUMAN EYE

POLARIZATION

To talk about the polarization of an electromagnetic wave, it's easiest to look at polarized light. Just remember that whatever applies to light generally applies to other forms of electromagnetic waves, too. So, what is meant by polarized light? It's light in which there's a preferred direction for the electric and magnetic field vectors in the wave. In unpolarized light, there is no preferred direction: the waves come in with electric and magnetic field vectors in random directions. In linearly polarized light, the electric field vectors are all along one line (and so are the magnetic field vectors, because they're perpendicular to the electric field vectors). Most light sources emit unpolarized light, but there are several ways light can be polarized.

One way to polarize light is by reflection. Light reflecting off a surface will tend to be polarized, with the direction of polarization (the way the electric field vectors point) being parallel to the plane of the interface.

Another way to polarize light is by selectively absorbing light with electric field vectors pointing in a particular direction. Certain materials, known as dichroic materials, do this, absorbing light polarized one way but not absorbing light polarized perpendicular to that direction. If the material is thick enough to absorb all the light polarized in one direction, the light emerging from the material will be linearly polarized. Polarizers (such as the lenses of polarizing sunglasses) are made from this kind of material.

If unpolarized light passes through a polarizer, the intensity of the transmitted light will be 1/2 of what it was coming in. If linearly polarized light passes through a polarizer, the intensity of the light transmitted is given by Malus' law:

Malus' law: $S = S_0 \cos^2\theta$ where θ is the angle between the direction of polarization of the incident light and the polarization axis of the polarizer.

A third way to polarize light is by scattering. Light scattering off atoms and molecules in the atmosphere is unpolarized if the light keeps traveling in the same direction, is linearly polarized if at scatters in a direction perpendicular to the way it was traveling, and somewhere between linearly polarized and unpolarized if it scatters of at another angle.

There are plenty of materials that affect the polarization of light. Certain materials (such as calcite) exhibit a property known as birefringence. A crystal of birefringent material affects light polarized in a particular direction differently from light polarized at 90 degrees to that direction; it refracts light polarized one way at a different angle than it refracts light polarized the other way. Looking through a birefringent crystal at something, you'd see a double image.

Liquid crystal displays, such as those in digital watches and calculators, also exploit the properties of polarized light.

Polarization by Reflection

One way to polarize light is by reflection. If a beam of light strikes an interface so that there is a 90° angle between the reflected and refracted beams, the reflected beam will be linearly polarized. The direction of polarization (the way the electric field vectors point)is parallel to the plane of the interface.

The special angle of incidence that satisfies this condition, where the reflected and refracted beams are perpendicular to each other, is known as the Brewster angle. The Brewster angle, the angle of incidence required to produce a linearly-polarized reflected beam, is given by:

$$\text{Brewster angle: } \tan\theta_B = n_2 / n_1$$

This expression can be derived using Snell's law, and the law of reflection. The diagram below shows some of the geometry involved.

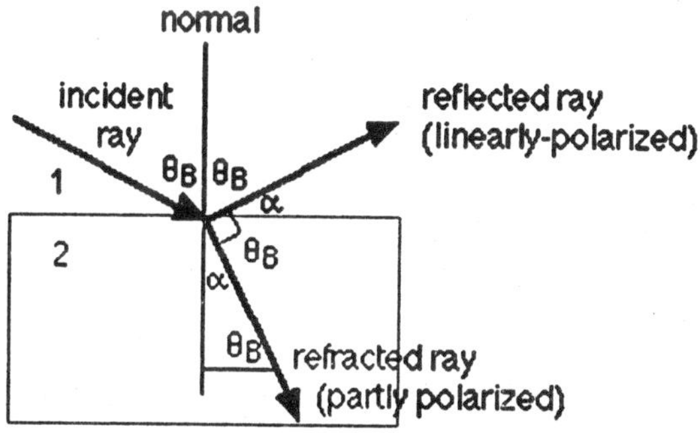

Using Snell's law:

$n_1 \sin\theta B = n_2 \sin\alpha = n_2 \cos\theta_B$

This gives the Brewster relationship: $\sin\theta_B/\cos\theta_B = n_2/n_1$

THE SCATTERING OF LIGHT IN THE ATMOSPHERE

The way light scatters off molecules in the atmosphere explains why the sky is blue and why the sun looks red at sunrise and sunset. In a nutshell, it's because the molecules scatter light at the blue end of the visible spectrum much more than light at the red end of the visible spectrum. This is because the scattering of light (i.e., the probability that light will interact with molecules when it passes through the atmosphere) is inversely proportional to the wavelength to the fourth power.

Scattering of light by molecules in the atmosphere is proportional to $1/\lambda^4$

Violet light, with a wavelength of about 400 nm, is almost 10 times as likely to be scattered than red light, which has a wavelength of about 700 nm. At noon, when the Sun is high in the sky, light from the Sun passes through a relatively thin layer of atmosphere, so only a small fraction of the light will be scattered. The Sun looks yellow-white because all the colours are represented almost equally. At sunrise or sunset, on the other hand, light from the Sun has to pass through much more atmosphere to reach our eyes. Along the way, most of the light towards the blue end of the spectrum is scattered in other directions, but much less of the light towards the red end of the spectrum is scattered, making the Sun appear to be orange or red.

So why is the sky blue? Again, let's look at it when the Sun is high in the sky. Some of the light from the Sun traveling towards other parts of the Earth is scattered towards us by the molecules in the atmosphere. Most of this scattered light is light from the blue end of the spectrum, so the sky appears blue.

Why can't this same argument be applied to clouds? Why do they look white, and not blue? It's because of the size of the water droplets in clouds. The droplets are much larger than the molecules in the atmosphere, and they scatter light of all colours equally. This makes them look white.

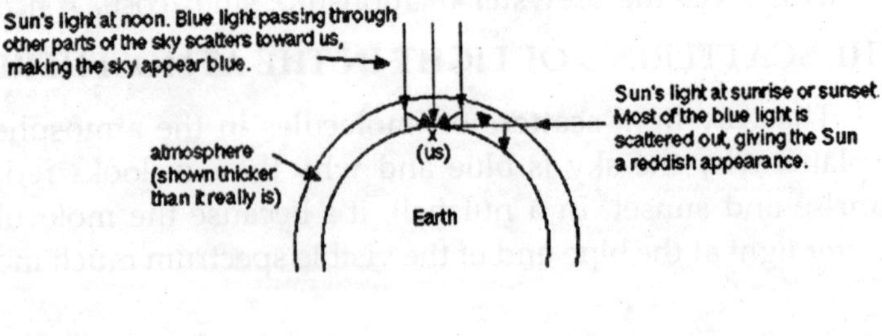

OPTICS OF THE EYE

The human eye is a wonderful instrument, relying on refraction and lenses to form images. There are many similarities between the human eye and a camera, including:

- A diaphragm to control the amount of light that gets through to the lens. This is the shutter in a camera, and the pupil, at the centre of the iris, in the human eye.
- A lens to focus the light and create an image. The image is real and inverted.
- A method of sensing the image. In a camera, film is used to record the image; in the eye, the image is focused on the retina, and a system of rods and cones is the front end of an image-processing system that converts the image to electrical impulses and sends the information along the optic nerve to the brain.

The way the eye focuses light is interesting, because most of the refraction that takes place is not done by the lens itself, but by the aqueous humour, a liquid on top of the lens. Light is refracted when it comes into the eye by this liquid, refracted a little more by the lens, and then a bit more by the vitreous humour, the jelly-like substance that fills the space between the lens and the retina.

The lens is critical in forming a sharp image, however; this is one of the most amazing features of the human eye, that it can adjust so quickly when focusing objects at different distances. This process of adjustment is known as accommodation.

Consider the lens equation:

$1/d_0 + 1/d_i = 1/f$ So $1/d_i = 1/f - 1/d_0$

With a camera, the lens has a fixed focal length. If the object distance is changed, the image distance (the distance between the lens and the film) is adjusted by moving the lens. This can't be done with the human eye: the image distance, the distance between the lens and the retina, is fixed. If the object distance is changed (i.e., the eye is trying to focus objects that are at different distances), then the focal length of the eye is adjusted to create a sharp image. This is done by changing the shape of the lens; a muscle known as the ciliary muscle does this job.

NEARSIGHTEDNESS

A person who is nearsighted can only create sharp images of close objects. Objects that are further away look fuzzy because the eye brings them in to focus at a point in front of the retina. To correct for this, a diverging lens is placed in front of the eye, diverging the light rays just enough so that when the rays are converged by the eye they converge on the retina, creating a focused image.

FARSIGHTEDNESS

A farsighted person can only create clear images of objects that are far away. Close objects are brought to a focus behind the retina, which is why they look fuzzy. To correct for this, a converging lens is placed in front of the eye, allowing images to be brought into sharp focus at the retina.

DIOPTERS

If you go to an optometrist to get glasses or contact lenses, you will get a prescription specified in units of diopters. This is a measure of the refractive power of the lens needed, which means it's a measure of the focal length of the lens. The two are, in fact, inversely related:

Refractive power in diopters = 1 / focal length in meters

A diopter has units of 1 / m.

If the lenses you get are specified as 5.0 diopters, it means they have a focal length of 0.2 m, meaning that they are converging lenses that bring parallel rays of light to a focus 0.2 m beyond the lens. Similarly, lenses of –2.0 diopters correspond to a focal length of –0.5 m; these would be diverging lenses with a focal point 0.5 m from the lens.

LENS ABERRATIONS

Lenses can distort the image of an object for a number of reasons. One kind of distortion is known as spherical aberration; this occurs because parallel rays of light are not all focused to a single point by a spherical lens. The further a ray is from the principal axis, the more it misses the focal point.

A second kind of distortion is chromatic aberration. This occurs because a lens will have a slightly different index of refraction for light of different wavelengths (i.e., light of different colours). In other words, chromatic aberration is caused by dispersion. Parallel rays of red light, therefore, would be brought to a different focal point than parallel rays of light of another colour, leading to a blurry image.

To minimize chromatic aberration, many high-quality lenses are made up of two lenses made from different materials. One lens will be converging, and the other diverging; the compound lens will still be converging overall, generally. The chromatic aberration introduced by one lens is corrected for in the second lens, bringing parallel rays of any colour light to about the same focal point.

OPTICAL INSTRUMENTS

When you use an optical instrument, whether it be something very simple like a magnifying glass, or more complicated like a telescope or microscope, you're usually trying to make things look bigger so you can more easily see fine details. One thing to remember about this is that if you want to make things look bigger, you're always going to use converging mirrors or lenses. Diverging mirrors or lenses always give smaller images.

When using a converging lens, it's helpful to remember these rules of thumb. If the object is very far away, the image will be tiny and very close to the focal point. As the object moves towards the lens, the image moves out from the focal point, growing as it does so. The object and image are exactly the same size when the object is at 2F, twice the focal distance from the lens. Moving the object from 2F towards F, the image keeps moving out away from the lens, and growing, until it goes to infinity when the object is at F, the focal point. Moving the object still closer to the lens, the image steadily comes in towards the lens from minus infinity, and gets smaller the closer the object is to the lens.

Note that similar rules of thumb apply for a converging mirror, too.

MULTIPLE LENSES

Many useful devices, such as microscopes and telescopes, use more than one lens to form images. To analyse any system with more than one lens, work in steps. Each lens takes an object and creates an image. The original object is the object for the first lens, and that creates an image. That image is the object for the second lens, and so on. We won't use more than two lenses, and we can do a couple of examples to see how you analyse problems like this.

A MICROSCOPE

A basic microscope is made up of two converging lenses. One reason for using two lenses rather than just one is that it's easier to get higher magnification. If you want an overall magnification of 35, for instance, you can use one lens to magnify by a factor of 5, and the second by a factor of 7. This is generally easier to do than to get magnification by a factor of 35 out of a single lens.

A microscope arrangement is shown below, along with the ray diagram showing how the first lens creates a real image. This image is the object for the second lens, and the image created by the second lens is the one you'd see when you looked through the microscope.

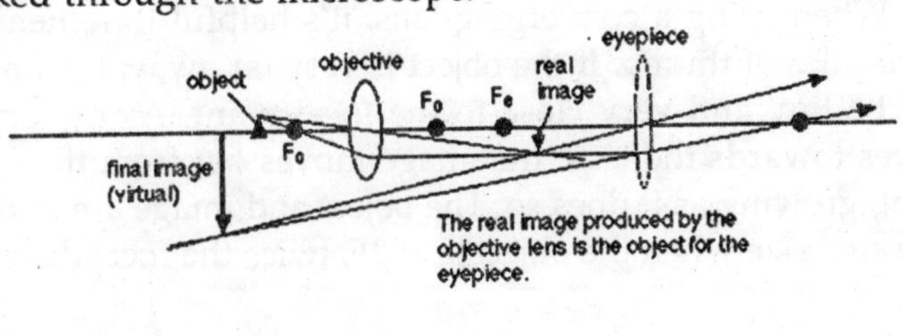

Note that the final image is virtual, and is inverted compared to the original object. This is true for many types of microscopes and telescopes, that the image produced is inverted compared to the object.

SIGN CONVENTION

The sign convention for lenses is similar to that for mirrors. Again, take the side of the lens where the object is to be the positive side. Because a lens transmits light rather than reflecting it like a mirror does, the other side of the lens is the positive side for images. In other words, if the image is on the far side of the lens as the object, the image distance is positive and the image is real. If the image and object are on the same side of the lens, the image distance is negative and the image is virtual.

For converging mirrors, the focal length is positive. Similarly, a converging lens always has a positive f, and a diverging lens has a negative f.

The signs associated with magnification also work the same way for lenses and mirrors. A positive magnification corresponds to an upright image, while a negative magnification corresponds to an inverted image. As usual, upright and inverted are taken relative to the orientation of the object.

Note that in certain cases involving more than one lens the object distance can be negative. This occurs when the image from the first lens lies on the far side of the second lens; that image is the object for the second lens, and is called a virtual object.

THE MICROSCOPE

Let's use the ray diagram for the microscope and work out a numerical example. The parameters we need to specify are:

d_0 = 7.0mm h_0 = 1.0mm

f_0 = 4.5mm f_e = 10.0mm

distance between the two lenses = 20mm

To work out the image distance for the image formed by the objective lens, use the lens equation, rearranged to:

$d_i = d_0 f/(d_0 - f) = (7)(4.5)/(2.5) = 12.6\text{mm}$

The magnification of the image in the objective lens is:

$m_0 = -d_i/d_0 = -12.6/7.0 = -1.8$

So the height of the image is -1.8 × 1.0 = -1.8 mm.

This image is the object for the second lens, and the object distance has to be calculated:

$$d_o = 20 - 12.6 = 7.4 \text{ mm}$$

The image, virtual in this case, is located at a distance of:

$$d_i = d_o f / (d_o - f) = (7.4)(10)/(-2.6) = -28.5 \text{ mm}$$

The magnification for the eyepiece is:

$m_e = - d_i / d_o = 28.5 / 7.4 = 3.85$

So the height of the final image is -1.8 mm x 3.85 = -6.9 mm.

The overall magnification of the two lens system is:

$m = m_0 \times m_e = -1.8 \times 3.85 = -6.9$

This is equal to the final height divided by the height of the object, as it should be. Note that, applying the sign conventions, the final image is virtual, and inverted compared to the object. This is consistent with the ray diagram.

TELESCOPES

A telescope needs at least two lenses. This is because you use a telescope to look at an object very far away, so the first lens creates a small image close to its focal point. The telescope is designed so the real, inverted image created by the first lens is just a little closer to the second lens than its focal length. As with the magnifying glass, this gives a magnified virtual image. This final image is also inverted compared to the original object. With astronomical telescopes, this doesn't really matter, but if you're looking at something on the Earth you generally want an upright image. This can be obtained with a third lens.

Note that the overall effect of the telescope is to magnify, which means the absolute value of the magnification must be larger than 1. The first lens (the objective) has a magnification smaller than one, so the second lens (the eyepiece) must magnify by a larger factor than the first lens reduces by. To a good approximation, the overall magnification is equal to the ratio of the focal lengths. With o standing for objective and e for eyepiece, the magnification is given by:

$m = - f_o / f_e$, with the minus sign meaning that the image is inverted.

RESOLVING POWER

The resolving power of an optical instrument, such as your eye, or a telescope, is its ability to separate far-away objects that are close together into individual images, as opposed to a single merged image. If you look at two stars in the sky, for example, you can tell they are two stars if they're separated by a large enough angle. Some stars, however, are so close together that they look like one star. You can only see that they are two stars by looking at them through a telescope. So, why does the telescope resolve the stars into separate objects while your eye can not? It's all because of diffraction.

If you look at a far-away object, the image of the object will form a diffraction pattern on your retina. For two far-away objects separated by a small angle, the diffraction patterns will overlap. You are able to resolve the two objects as long as the central peaks in the two diffraction patterns don't overlap. The limit is when one central peak falls at the position of the first dark fringe for the second diffraction pattern. This is known as the Rayleigh criterion. Once the two central peaks start to overlap, in other words, the two objects look like one.

The size of the central peak in the diffraction pattern depends on the size of the aperture (the opening you look through). For your eye, this is your pupil. A telescope, or even a camera, has a much larger aperture, and therefore more resolving power. The minimum angular separation is given by:

$\theta_{min} = 1.22\lambda/D$ where D is the size of the aperture.

The factor of 1.22 applies to circular apertures like your pupil, a telescope, or a camera lens.

The closer you are to two objects, the greater the angular separation between them. Up close, then, two objects are easily resolved. As you get further from the objects, however, they will eventually merge to become one.

X-RAY DIFFRACTION

Things that look a lot like diffraction gratings, orderly arrays of equally-spaced objects, are found in nature; these are crystals. Many solid materials (salt, diamond, graphite, metals, etc.) have a crystal structure, in which the atoms are arranged in a repeating, orderly, 3-dimensional pattern. This is a lot like a diffraction grating, only a three-dimensional grating. Atoms in a typical solid are separated by an angstrom or a few angstroms;. This is much smaller than the wavelength of visible light, but x-rays have wavelengths of about this size. X-rays interact with crystals, then, in a way very similar to the way light interacts with a grating.

X-ray diffraction is a very powerful tool used to study crystal structure. By examining the x-ray diffraction pattern, the type of crystal structure (i.e., the pattern in which the atoms are arranged) can be identified, and the spacing between atoms can be determined.

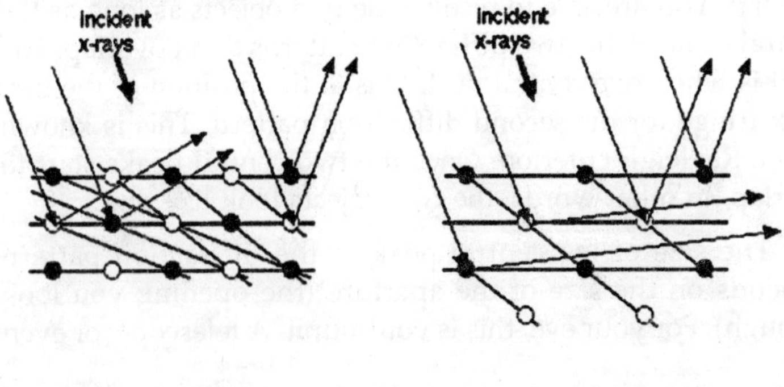

The two diagrams below can help to understand how x-ray diffraction works. Each represents atoms arranged in a particular crystal structure.

You can think of the diffraction pattern like this. When x-rays come in at a particular angle, they reflect off the different planes of atoms as if they were plane mirrors. However, for a particular set of planes, the reflected waves interfere with each other. A reflected x-ray signal is only observed if the conditions are right for constructive interference. If d is the distance between planes, reflected x-rays are only observed under these conditions:

$2d \sin\theta = m\lambda$ (θ is the angle between the normal to the plane and the incident (and reflected) x-rays)

That's known as Bragg's law. The important thing to notice is that the angles at which you see reflected x-rays are related to the spacing between planes of atoms. By measuring the angles at which you see reflected x-rays, you can deduce the spacing between planes and determine the structure of the crystal.

ELECTRONS, PHOTONS, AND THE PHOTO-ELECTRIC EFFECT

We're now starting to talk about quantum mechanics, the physics of the very small.

PLANCK'S CONSTANT

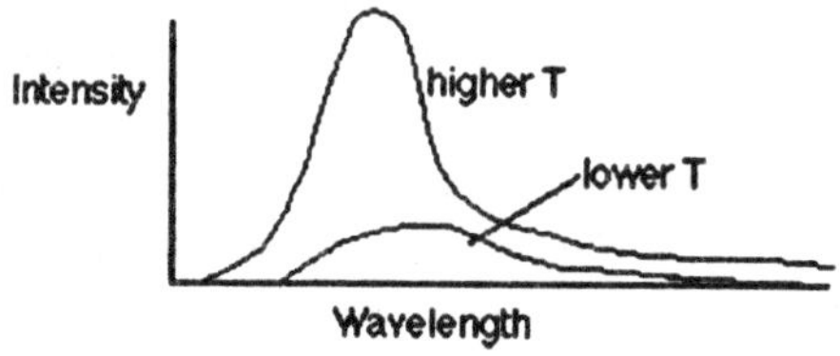

The spectrum of radiation emitted by a blackbody at different temperatures.

At the end of the 19th century one of the most intriguing puzzles in physics involved the spectrum of radiation emitted by a hot object. Specifically, the emitter was assumed to be a blackbody, a perfect radiator. The hotter a blackbody is, the more the peak in the spectrum of emitted radiation shifts to shorter wavelength. Nobody could explain why there was a peak in the distribution at all, however; the theory at the time predicted that for a blackbody, the intensity of radiation just kept increasing as the wavelength decreased. This was known as the ultraviolet catastrophe, because the theory predicted that an infinite amount of energy was emitted by a radiating object.

Clearly, this prediction was in conflict with the idea of conservation of energy, not to mention being in serious disagreement with experimental observation. No one could account for the discrepancy, however, until Max Planck came up with the idea that a blackbody was made up of a whole bunch of oscillating atoms, and that the energy of each oscillating atom was quantized. That last point is the key: the energy of the atoms could only take on discrete values, and these values depended on the frequency of the oscillation:

Planck's prediction of the energy of an oscillating atom: $E = nhf$ ($n = 0, 1, 2, 3...$)

where f is the frequency, n is an integer, and h is a constant known as Planck's constant. This constant shows up in many different areas of quantum mechanics.

$$\text{Planck's constant}: h = 6.63 \times 10^{-34} \text{ J s}$$

The spectra predicted for a radiating blackbody made up of these oscillating atoms agrees very well with experimentally-determined spectra.

Planck's idea of discrete energy levels led Einstein to the idea that electromagnetic waves have a particle nature. When Planck's oscillating atoms lose energy, they can do so only by making a jump down to a lower energy level. The energy lost by the atoms is given off as an electromagnetic wave. Because the energy levels of the oscillating atoms are separated by hf, the energy carried off by the electromagnetic wave must be hf.

THE PHOTOELECTRIC EFFECT

Einstein won the Nobel Prize for Physics not for his work on relativity, but for explaining the photoelectric effect. He proposed that light is made up of packets of energy called photons. Photons have no mass, but they have momentum and they have an energy given by:

Energy of a photon: $E = hf$

The photoelectric effect works like this. If you shine light of high enough energy on to a metal, electrons will be emitted from the metal. Light below a certain threshold frequency, no matter how intense, will not cause any electrons to be emitted. Light above the threshold frequency, even if it's not very intense, will always cause electrons to be emitted.

The explanation for the photoelectric effect goes like this: it takes a certain energy to eject an electron from a metal surface. This energy is known as the work function (W), which depends on the metal. Electrons can gain energy by interacting with photons. If a photon has an energy at least as big as the work function, the photon energy can be transferred to the electron and the electron will have enough energy to escape from the metal. A photon with an energy less than the work function will never be able to eject electrons.

Before Einstein's explanation, the photoelectric effect was a real mystery. Scientists couldn't really understand why low-frequency high-intensity light would not cause electrons to be emitted, while higher-frequency low-intensity light would. Knowing that light is made up of photons, it's easy to explain now. It's not the total amount of energy (i.e., the intensity) that's important, but the energy per photon.

When light of frequency f is incident on a metal surface that has a work function W, the maximum kinetic energy of the emitted electrons is given by:

$KE_{max} = hf - W$

Note that this is the maximum possible kinetic energy because W is the minimum energy necessary to liberate an electron. The threshold frequency, the minimum frequency the photons can have to produce the emission of electrons, is when the photon energy is just equal to the work function:

Threshold frequency: $f_0 = W/h$

THE COMPTON EFFECT

Although photons have no mass, they do have momentum, given by:

momentum of a photon: $p = E/c = hf/c = h/\lambda$

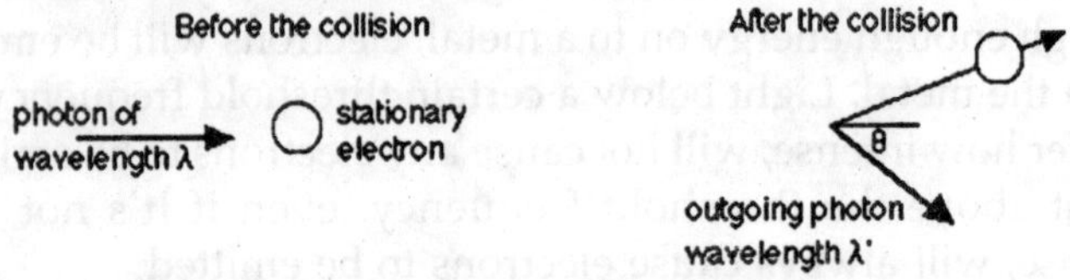

Convincing evidence for the fact that photons have momentum can be seen when a photon collides with a stationary electron. Some of the energy and momentum is transferred to the electron (this is known as the Compton effect), but both energy and momentum are conserved in such a collision. Applying the principles of conservation of energy and momentum to this collision, one can show that the wavelength of the outgoing photon is related to the wavelength of the incident photon by the equation:

Compton effect: $\lambda' - 1\lambda = (h/mc)[1\text{-}\cos\theta]$

where θ is the angle between the incident and autgoing phtons, and m is the mass of the electron.

WAVE-PARTICLE DUALITY

To explain some aspects of light behaviour, such as interference and diffraction, you treat it as a wave, and to explain other aspects you treat light as being made up of particles. Light exhibits wave-particle duality, because it exhibits properties of both waves and particles.

Wave-particle duality is not confined to light, however. Everything exhibits wave-particle duality, everything from electrons to baseballs. The behaviour of relatively large objects, like baseballs, is dominated by their particle nature; to explain the behaviour of very small things like electrons, both the wave properties and particle properties have to be considered.

Electrons, for example, exhibit the same kind of interference pattern as light does when they're incident on a double slit.

THE DE BROGLIE WAVELENGTH

Louis de Broglie predicted that since light exhibited both wave and particle behaviour, particles should also. He proposed that all particles have a wavelength given by:

de Broglie wavelength: $\lambda = h/p$

Note that this is the same equation that applies to photons.

de Broglie's prediction was shown to be true when beams of electrons and neutrons were directed at crystals and diffraction patterns were seen. This is evidence of the wave properties of these particles.

Everything has a wavelength, but the wave properties of matter are only observable for very small objects. If you work out the wavelength of a moving baseball, for instance, you will find that the wavelength is far too small to be observable.

What is a Particle Wave?

The probability of finding a particle at a particular location, then, is related to the wave associated with the particle. The larger the amplitude of the wave at a particular point, the larger the probability that the electron will be found there. Similarly, the smaller the amplitude the smaller the probability. In fact, the probability is proportional to the square of the amplitude of the wave.

QUANTUM MECHANICS

All these ideas, that for very small particles both particle and wave properties are important, and that particle energies are quantized, only taking on discrete values, are the cornerstones of quantum mechanics. In quantum mechanics we often talk about the wave function ψ of a particle; the wave function is the wave discussed above, with the probability of finding the particle in a particular location being proportional to the square of the amplitude of the wave function.

UNDERSTANDING THE ATOM

RUTHERFORD SCATTERING

Let's focus on the atom, starting from a historical perspective. Ernest Rutherford did a wonderful experiment in which he fired alpha particles (basically helium nuclei) at a very thin gold foil. He got a rather surprising result: rather than all the particle passing straight through the foil, many were scattered off at large angles, some even coming straight back. This was inconsistent with the plum-pudding model of the atom, in which the atom was viewed as tiny electrons embedded in a dispersed pudding of positive charge. Rutherford proposed that the positive charge must really be localized, concentrated in a small nucleus.

This led to the planetary model of the atom, with electrons orbiting the nucleus like planets orbiting the Sun.

LINE SPECTRA

If the atom looked like a solar system, how could line spectra be explained? Line spectra are what you get when you excite gases with a high voltage. Gases emit light at only a few sharply-defined frequencies, and the frequencies are different for different gases. These emission spectra, then, are made up of a few well-defined lines.

Gases will also selectively absorb light at these same frequencies. You can see this if you expose a gas to a continuous spectrum of light. The absorption spectra will be very similar to a continuous spectrum, except for a few dark lines corresponding to the frequencies absorbed by the gas.

THE BOHR MODEL

The Bohr model is a planetary model of the atom that explains things like line spectra. Neils Bohr proposed that the electrons orbiting the atom could only occupy certain orbits, orbits in which the angular momentum satisfied a particular equation:

$$L_n = mv_nr_n = nh/2\pi \ (n = 1, 2, 3,...)$$

where m is the mass of the electron, r is the radius of the orbit, and v is the orbital speed of the electron.

In other words, Bohr was proposing that the angular momentum of an electron in an atom is quantized.

What does quantization of the angular momentum mean for the energy of the electron in a particular orbit? We can analyse the energy very simply using concepts of circular motion and the potential energy associated with two charges. The electron has a charge of -e, while the nucleus has a charge of +Ze, where Z is the atomic number of the element. The energy is then given by:

Energy of an electron: $E = KE + PE = 1/2\ mv^2 - kZe^2/r$

The electron is experiencing uniform circular motion, with the only force on it being the attractive force between the negative electron and the positive nucleus. Thus:

$mv^2/r = kze^2/r^2$ so $KE = 1/2\ mv^2 = 1/2\ kZe^2/r$

Plugging this back into the energy equation gives:

Energy of an electron: $E = -1/2\ kZe^2/r$

If you rearrange the angular momentum equation to solve for the velocity, and then plug that back into the equation:

$mv^2 = kZe^2/r$

and solve that for r, you get:

$$r_n = [\,h^2/(4\pi^2 mke^2)]\,n^2/Z = 5.29 \times 10^{-11}\,n^2/Z \qquad (n = 1, 2, 3, ...)$$

This can now be substituted into the energy equation, giving the total energy of the nth level:

$$\text{Bohr energy}: E_n = -(2\pi^2 mk^2e^4/h^2)\,Z^2/n^2 = -(13.6\ \text{eV})\ Z^2/n^2 \quad (n = 1, 2, 3, ...)$$

ENERGY LEVEL DIAGRAMS AND THE HYDROGEN ATOM

It's often helpful to draw a diagram showing the energy levels for the particular element you're interested in. The diagram for hydrogen is shown on page 918 in the text. Hydrogen's easy to deal with because there's only one electron to worry about.

The n = 1 state is known as the ground state, while higher n states are known as excited states. If the electron in the atom makes a transition from a particular state to a lower state, it is losing energy. To conserve energy, a photon with an energy equal to the energy difference between the states will be emitted by the atom. In the hydrogen atom, with Z = 1, the energy of the emitted photon can be found using:

photon energy: $E = hf = (13.6eV)\ (1/nf^2 - 1/ni^2)$

where nf is the n of the final state and n_i is the n of the initial state.

Atoms can also absorb photons. If a photon with an energy equal to the energy difference between two levels is incident on an atom, the photon can be absorbed, raising the electron up to the higher level.

ANGULAR MOMENTUM

Bohr's model of the atom was based on the idea the angular momentum is quantized, and quantized in a particular way. de Broglie came up with an explanation for why the angular momentum might be quantized in this way. de Broglie realised that if you use the wavelength associated with the electron, and only allow for standing waves to exist in any orbit (in other words, the circumference of the orbit has to be an integral number of wavelengths), then you arrive at the same relationship for the angular momentum that Bohr got.

The derivation works like this, starting from the idea that the circumference of the circular orbit must be an integral number of wavelengths:

$$2\pi r = n\lambda$$

Taking the wavelength to be the de Broglie wavelength, this becomes:

$$\lambda = h/p \qquad \text{so } 2\pi r = nh/p$$

The momentum, p, is simply mv as long as we're talking about non-relativistic speeds, so this becomes:

$$2\pi r = nh/mv$$

Rearranging this a little, and recognizing that the angular momentum for a point mass is simply L = mvr, gives the Bohr relationship:

$$mvr = L = nh/2\pi$$

THE QUANTUM MECHANICAL VIEW OF THE ATOM

HEISENBERG UNCERTAINTY PRINCIPLE

The uncertainty principle is a rather interesting idea, stating that it is not possible to measure both the position and momentum of a particle with infinite precision. It also states that the more accurately you measure a particle's position, the less accurately you're able to measure it's momentum, and vice versa.

This idea is really not relevant when you're making measurements of large objects. It is relevant, however, when you're looking at very small objects such as electrons. Consider that you're trying to measure the position of an electron. To do so, you bounce photons off the electron; by figuring out the time it takes for each photon to come back to you, you can figure out where the electron is. The more photons you use, the more precisely you can measure the electron's position.

However, each time a photon bounces off the electron, momentum is transferred to the electron. The more photons you use, the more momentum is transferred, and because you can't measure that momentum transferred to infinite precision the more uncertainty you're introducing in the measurement of the momentum of the electron.

Heisenberg showed that there is a limit to the accuracy you can measure things:

Heisenberg uncerainty principle: $\Delta y \Delta p \geq h/2\pi$

where Δy is the uncertainty in the measured position, and Δp is the uncertainty in the momentum.

The uncertainty can also be stated in terms of the energy of a particle in a particular state, and the time in which the particle is in that state:

Heisenberg unceraintу principle: $\Delta E \Delta t \geq h/2\pi$

where Δt is the time interval during which the particle is in a state with energy E.

QUANTUM NUMBERS

The Bohr model of the atom involves a single quantum number, the integer n that appears in the expression for the energy of an electron in an orbit. This picture of electrons orbiting a nucleus in well-defined orbits, the way planets orbit the Sun, is not our modern view of the atom. We now picture the nucleus surrounded by electron clouds, so the orbitals are not at all well-defined; we still find the Bohr theory to be useful, however, because it gives the right answer for the energy of the electron orbitals.

The Bohr model uses one quantum number, but a full quantum mechanical treatment requires four quantum numbers to characterize the electron orbitals. These are known as the principal quantum number, the orbital quantum number, the magnetic quantum number, and the spin quantum number. These are all associated with particular physical properties.

n, the principal quantum number, is associated with the total energy, the same way it is in the Bohr model. In fact, calculating the energy from the quantum mechanical wave function gives the expression Bohr derived for the energy:

Energy: $E_n = -(13.6\text{eV})Z^2/n^2$ (n=1, 2, 3,....)

e, the orbital quantum number, is connected to the total angular momentum of the electron. This quantum number is an integer less than n, and the total angular momentum of the electron can be calculated using:

Total angular momentum: $L = [e(e + 1)]^{1/2}h/2\pi$ (e = 0, 1, 2,..., n – 1) *me*, the magnetic quantum number, is related to one particular component of the angular momentum. By convention, we call this the z-component. The energy of any orbital depends on the magnetic quantum number only when the atom is in an external magnetic field. This quantum number is also an integer; it can be positive or negative, but it has a magnitude less than or equal to the orbital quantum number. The z-component of the electron's angular momentum is given by:

$L_Z = meh/2\pi$ (me = –e,..., –2, –1, 0, 1, 2,..., e)

m_s, the spin quantum number is related to something called the spin angular momentum of the electron. The closest analogy is that it's similar to the Earth spinning on its axis. There are only two possible states for this quantum number, often referred to as spin up and spin down.

$m_s = +1/2$ or $m_s = -1/2$

What's the use of having all these quantum numbers? We need all four to completely describe the state an electron occupies in the atom.

ELECTRON PROBABILITY DENSITY CLOUDS

A very important difference between the Bohr model and the full quantum mechanical treatment of the atom is that Bohr proposed that the electrons were found in very well-defined circular orbits around the nucleus, while the quantum mechanical picture of the atom has the electron essentially spread out into a cloud. We call this a probability density cloud, because the density of the cloud tells us what the probability is of finding the electron at a particular distance from the nucleus.

In quantum mechanics, something called a wave function is associated with each electron state in an atom. The probability of finding an electron at a particular distance from the nucleus is related to the square of the wave function, so these electron probability density clouds are basically three-dimensional pictures of the square of the wave function.

THE PAULI EXCLUSION PRINCIPLE

If you've got a hydrogen atom, with only a single electron, it's very easy to determine the possible states that electron can occupy. A particular state means one particular combination of the 4 quantum numbers; there are an infinite number of states available, but the electron is more likely to occupy a low-energy state (i.e., a low n state) than a higher-energy (higher n) state.

What happens for other elements, when there is more than one electron to worry about? Can all the electrons be found in one state, the ground state, for example? It turns out that this is forbidden: the Pauli exclusion principle states that no two electrons can occupy the same state. In other words, no two electrons can have the same set of 4 quantum numbers.

SHELLS AND SUBSHELLS

As usual, for historical reasons we have more than one way to characterize an electron state in an atom. We can do it using the 4 quantum numbers, or we can use the notion of shells and subshells. A shell consists of all those states with the same value of n, the principal quantum number. A subshell groups all the states within one shell with the same value of *e*, the orbital quantum number.

The subshells are usually referred to by letters, rather than by the corresponding value of the orbital quantum number. The letters s, p, d, f, g, and h stand for values of 0, 1, 2, 3, 4, and 5, respectively. Using these letters allows us to use a shorthand to denote how many electrons are in a subshell; this is useful for specifying the ground state (lowest energy state) of a particular atom.

The ground state configuration for oxygen, for instance, can be written as:

oxygen: $1s^2 2s^2 2s^4$

This means that the lowest energy configuration of oxygen, with 8 electrons, is to have two electrons in the n=1 s-subshell, two in the n=2 s-subshell, and four in the n=2 p-subshell.

Potassium (Z = 19) has an interesting ground state configuration:

$$\text{potassium}: 1s^2\ 2s^2\ 2p^6\ 3s^2\ 3p^6\ 4s^1$$

That's interesting because there is a d-subshell in the n = 3 shell, but instead of the last electron going into that subshell it goes into the s-subshell of the n=4 shell. It does this to minimize the energy: the 4s subshell is at a lower energy than the 3d subshell.

THE PERIODIC TABLE

When Mendeleev organized the elements into the periodic table, he knew nothing about quantum numbers and subshells. The way the elements are organized in the periodic table, however, is directly related to how the electrons fill the levels in the different shells.

Different columns of the periodic table group elements with similar properties; they have similar properties because of the similarities between their ground state electron configurations. The noble gases (He, Ne, Ar, etc.) are all in the right-most column of the periodic table. Their ground state configurations have no partially filled subshells; having a complete subshell is favorable from the standpoint of minimizing energy so these elements do not react readily.

On the other hand, the column next to the noble gases is the halogens; these are one electron short of having completely-filled subshells, so if they can share an electron from another element they're happy to do so. They react readily with elements whose ground state configurations have a single electron in one subshell like the alkali metals (Li, Na, K, etc.).

Chapter 3

Light is a Wave or a Particle

LIGHT IS A PARTICLE

- It travels in straight lines;
- It forms sharp shadows;
- Light beams are localized in space;
- It was easier to imagine a stream of particles traveling in a vacuum

LIGHT IS A WAVE

Two crossing beams do not collide,

Light beams travel at only one speed

MAXWELL'S EQUATIONS

Maxwell develops electromagnetic wave theory of light. He shows that moving charges can generate a wave that travels with exactly the measured speed of light.

But – the question arises again in the 20th century in connection with black body radiation and quantum mechanics. The wave model of Maxwell is found to be inadequate – especially when describing the interaction between light and matter.Furthermore, classical particles can also exhibit wave-like behaviour under some circumstances. The wavelength associated with a particle is usually very small, so that l/D for a particle is almost always much less than 1.Both classical "particles" and classical "waves" share common characteristics and the distinction is not always clear. The parameter l/D still counts.

In the latter half of the nineteenth century, the Scottish physicist James Clerk Maxwell demonstrated that all previously established experimental facts regarding electric and magnetic fields could be summed up in just *four* equations. Nowadays, these equations are generally known as *Maxwell's equations*.

The first equation is simply Gauss' law. This equation describes how electric charges generate electric fields. Gauss' law states that:

The electric flux through any closed surface is equal to the total charge enclosed by the surface, divided by.

This can be written mathematically as

$$\oint_S \mathbf{E} \cdot d\mathbf{S} = \frac{Q}{\epsilon_0},$$

where S is a closed surface enclosing the charge Q. The above expression can also be written

$$\oint_S \mathbf{E} \cdot d\mathbf{S} = \frac{1}{\epsilon_0} \int_V \rho \, dV,$$

where V is a volume bounded by the surface S, ρ and is the *charge density*: *i.e.*, the electric charge per unit volume.

The second equation is the magnetic equivalent of Gauss' law. This equation describes how the non-existence of magnetic monopoles causes magnetic field-lines to form closed loops. Gauss' law for magnetic fields states that:

The magnetic flux through any closed surface is equal to zero.

This can be written mathematically as

$$\int_S \mathbf{B} \cdot d\mathbf{S} = 0,$$

where S is a closed surface.

The third equation is Faraday's law. This equation describes how changing magnetic fields generate electric fields. Faraday's law states that:

The line integral of the electric field around any closed loop is equal to minus the time rate of change of the magnetic flux through the loop.

This can be written mathematically as

$$\oint_C \mathbf{E} \cdot d\mathbf{S} = -\frac{d}{dt}\int_{S'} \mathbf{B} \cdot d\mathbf{S}';$$

where S' is a surface attached to the loop C.

The fourth, and final, equation is Ampère's circuital law This equation describes how electric currents generates magnetic fields. Ampère's circuital law states that:

The line integral of the magnetic field around any closed loop is equal to times the algebraic sum of the currents which pass through the loop.

This can be written mathematically as

$$\oint_C \mathbf{B} \cdot d\mathbf{r} = \mu_0 I;$$

where I is the net current flowing through loop C. This equation can also be written

$$\oint_C \mathbf{B} \cdot d\mathbf{r} = \mu_0 \int_{S'} \mathbf{j} \cdot d\mathbf{S}';$$

where S' is a surface attached to the loop C, and J is the *current density*: *i.e.*, the electrical current per unit area.

When Maxwell first wrote Equations. He was basically trying to summarize everything which was known at the time about electric and magnetic fields in mathematical form. However, the more Maxwell looked at his equations, the more convinced he became that they were incomplete. Eventually, he proposed adding a new term, called the *displacement current*, to the right-hand side of his fourth equation. In fact, Maxwell was able to show that equations are *mathematically inconsistent* unless the displacement current term is added to Eq. Unfortunately, Maxwell's demonstration of this fact requires some advanced mathematical techniques which lie well beyond the scope of this course. In the following, we shall give a highly simplified version of his derivation of the missing term.

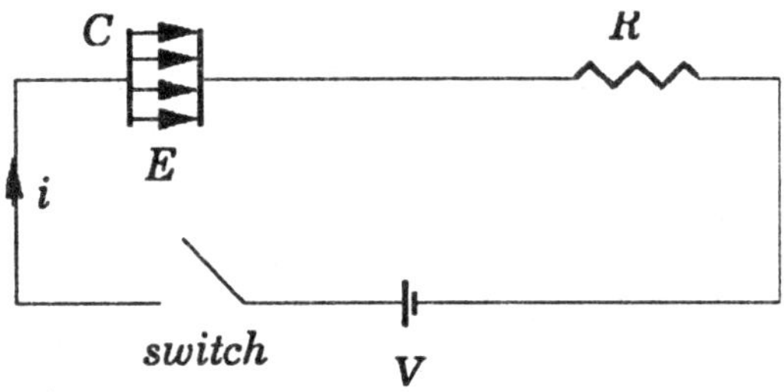

Fig. Circuit containing a charging capacitor.

Consider a circuit consisting of a parallel plate capacitor of capacitance C in series with a resistance R and an steady emf V, as shown in Fig. Let A be the area of the capacitor plates, and let d be their separation. Suppose that the switch is closed at $t = 0$. The current i flowing around the circuit starts from an initial value of $I = V/R$, and gradually decays to zero on the RC time of the circuit. Simultaneously, the charge q on the positive plates of the capacitor starts from zero, and gradually increases to a final value of $Q = CV$. As the charge q varies, so does the potential difference between the capacitor plates, since $v = q/C$.

The electric field in the region between the plates is approximately uniform, directed perpendicular to the plates (running from the positively charged plate to the negatively charged plate), and is of magnitude $E = v/d$. It follows that

$$q = Cv = C\,d\,E.$$

In a time interval dt, the charge on the positive plate of the capacitor increases by an amount $dq = CddE$, where dE is the corresponding increase in the electric field-strength between the plates. Note that both c and d are time-independent quantities. It follows that

$$\frac{dq}{dt} = C\,d\,\frac{dE}{dt}.$$

Now, dq/dt is numerically equal to the instantaneous current i flowing around the circuit (since all of the charge which flows around the circuit must accumulate on the plates of the capacitor). Also, $C = \varepsilon_0 A/d$ for a parallel plate capacitor. Hence, we can write

$$i = \frac{dq}{dt} = C\,d\frac{dE}{dt} = \epsilon_0 A \frac{dE}{dt}.$$

Since the electric field E is normal to the area A, we can also write

$$i = \epsilon_0 A \frac{dE_\perp}{dt}.$$

Equation relates the instantaneous current flowing around the circuit to the time rate of change of the electric field between the capacitor plates. According to Eq. the current flowing around the circuit generates a magnetic field. This field circulates around the current carrying wires connecting the various components of the circuit. However, since there is no actual current flowing between the plates of the capacitor, no magnetic field is generated in this region, according to Eq. Maxwell demonstrated that for reasons of mathematical self-consistency there must, in fact, be a magnetic field generated in the region between the plates of the capacitor. Furthermore, this magnetic field must be the same as that which would be generated if the current (*i.e.*, the same current as that which flows around the rest of the circuit) flowed between the plates. Of course, there is no actual current flowing between the plates. However, there is a changing electric field. Maxwell argued that a changing electric field constitutes an effective current (*i.e.*, it generates a magnetic field in just the same manner as an actual current). For historical reasons (which do not particularly interest us at the moment), Maxwell called this type of current a *displacement current*. Since the displacement current flowing between the plates of the capacitor must equal the current I_D flowing around the rest of the circuit, it follows from Equation. that

$$I_D = \epsilon_0 A \frac{dE_\perp}{dt}.$$

Above equation was derived for the special case of the changing electric field generated in the region between the plates of a charging parallel plate capacitor. Nevertheless, this equation turns out to be completely general. Note that $AE\perp$ is equal to the electric flux ϕE between the plates of the capacitor. Thus, the most general expression for the displacement current passing through some closed loop is

$$I_D = \varepsilon_0 \frac{d\phi E}{dt},$$

where ϕE is the electric flux through the loop.

According to Maxwell's argument, a displacement current is just as effective at generating a magnetic field as a real current. Thus, we need to modify Ampère's circuital law to take displacement currents into account. The modified law, which is known as the *Ampère-Maxwell law,* is written

The line integral of the electric field around any closed loop is equal to μ_0 times the algebraic sum of the actual currents and which pass through the loop plus μ_0 times the displacement current passing through the loop.

This can be written mathematically as

$$\oint_C B \cdot dr = \mu_0 \left(I + I_D\right),$$

where C is a loop through which the electric current I and the displacement current I_D pass. This equation can also be written

$$\oint_C B \cdot dr = \mu_0 \int_{S'} j \cdot dS' + \mu_0 \in_0 \frac{d}{dt} \int_{S'} E \cdot dS',$$

where S' is a surface attached to the loop C.

Equations are known collectively as *Maxwell's equations.* They constitute a complete and mathematically self-consistent description of the behaviour of electric and magnetic fields.

ELECTROMAGNETIC WAVES

One of the first things that Maxwell did with his four equations, once he had obtained them, was to look for wave-like solutions. Maxwell knew that the wave-like solutions of the equations of gas dynamics correspond to sound waves, and the wave-like solutions of the equations of fluid dynamics correspond to gravity waves in water, so he reasoned that if his equations possessed wave-like solutions then these would correspond to a completely new type of wave, which he called an *electromagnetic wave.*

Maxwell was primarily interested in electromagnetic waves which can propagate through a vacuum (*i.e.*, a region containing no charges or currents). Now, in a vacuum, Maxwell's equations reduce to

$$\oint_S E \cdot dS = 0$$

$$\oint_S B \cdot dS = 0$$

$$\oint_C E \cdot dr = -\frac{d}{dt}\int_{S'} B \cdot dS',$$

$$\oint_C B \cdot dr = \mu_0 \epsilon_0 \frac{d}{dt}\int_{S'} E.dS',$$

where S is a closed surface, and S' a surface attached to some loop C. Note that, with the addition of the displacement current term on the right-hand side of Eq. these equations exhibit a nice symmetry between electric and magnetic fields. Unfortunately, Maxwell's mathematical proof that the above equations possess wave-like solutions lies well beyond the scope of this course. We can, nevertheless, still write down these solutions, and comment on them.

Consider a plane electromagnetic wave propagating along the Z-axis. According to Maxwell's calculations, the electric and magnetic fields associated with such a wave take the form

$$E_x = E_0 \cos\left[2\pi(z/\lambda - ft)\right],$$

$$E_y = B_0 \cos\left[2\pi(z/\lambda - ft)\right].$$

Note that the fields are periodic in both time and space. The oscillation frequency (in hertz) of the fields at a given point in space is f. The equation of a wave crest is

$$\frac{z}{\lambda} - ft = N$$

where N is an integer. It can be seen that the distance along the z-axis between successive wave crests is given by λ. This distance is conventionally termed the *wavelength*. Note that each wave crest *propagates* along the z-axis. In a time interval dt, the Nth wave crest moves a distance $dz = \lambda f dt$, according to Eq. Hence, the velocity $c = dz/dt=$ with which the wave propagates along the z-axis is given by

$$c = f\lambda.$$

Maxwell was able to establish that electromagnetic waves possess the following properties:

- The magnetic field oscillates *in phase* with the electric field. In other words, a wave maximum of the magnetic field always coincides with a wave maximum of the electric field in both time and space.
- The electric field is always perpendicular to the magnetic field, and both fields are directed at right-angles to the direction of propagation of the wave. In fact, the wave propagates in the direction

  $E \times B$
- Electromagnetic waves are clearly a type of *transverse wave*.
- For a z-directed wave, the electric field is free to oscillate in *any* direction which lies in the x–y plane. The direction in which the electric field oscillates is conventionally termed the direction of *polarization* of the wave. Thus, Eqs. represent a plane electromagnetic wave which propagates along the z-axis, and is polarized in the x-direction.
- The maximum amplitudes of the electric and the magnetic fields are related via $E_0 = cB_0$.
- There is no constraint on the possible frequency or wavelength of electromagnetic waves. However, the propagation velocity of electromagnetic waves is *fixed*, and takes the value

$$c = \frac{1}{\sqrt{\mu_0 \in_0}}.$$

According to above equations a changing magnetic field generates an electric field, and a changing electric field generates a magnetic field. Thus, we can think of the propagation of an electromagnetic field through a vacuum as due to a kind of "leap-frog" effect, in which a changing electric field generates a magnetic field, which, in turn, generates an electric field, and so on. Note that the displacement current term in Eq. plays a crucial role in the propagation of electromagnetic waves. Indeed, without this term, a changing electric field is incapable of generating a magnetic field, and so there can be no leap-frog effect. Electromagnetic waves have many properties in common with other types of wave (*e.g.*, sound waves). However, they are unique in one respect: *i.e.*, they are able to propagate through a vacuum. All other types of waves require some sort of medium through which to propagate.

Maxwell deduced that the speed of propagation of an electromagnetic wave through a vacuum is entirely determined by the constants μ_0 and $\in_0$. The former constant is related to the strength of the magnetic field generated by a steady current, whereas the latter constant is related to the strength of the electric field generated by a stationary charge. The values of both constants were well known in Maxwell's day. In modern units, $\mu_0 = 4\pi \times 10^{-7} Ns^2C^{-2}$ and $\in_0 = 8.854 \times 10^{-12} C^2N^{-1}m^{-2}$. Thus, when Maxwell calculated the velocity of electromagnetic waves he obtained

$$c = \frac{1}{\sqrt{(4\pi \times 10^{-7})(8.854 \times 10^{-12})}} = 2.998 \times 10^8 \, \mathrm{m\,s^{-1}}.$$

Now, Maxwell knew [from the work of Fizeau (1849) and Foucault (1850)] that the velocity of light was about $3 \times 10^8 \mathrm{ms}^{-1}$. The remarkable agreement between this experimentally determined velocity and his theoretical prediction for the velocity of electromagnetic waves immediately lead Maxwell to hypothesize that *light is a form of electromagnetic wave.* Of course, this hypothesis turned out to be correct. We can still appreciate that Maxwell's achievement in identifying light as a form of electromagnetic wave was quite remarkable. After all, his equations were derived from the results of bench-top laboratory experiments involving charges, batteries, coils, and currents, *etc.*, which apparently had nothing whatsoever to do with light.

Maxwell was able to make another remarkable prediction. The wavelength of light was well known in the late nineteenth century from studies of diffraction through slits, *etc.* Visible light actually occupies a surprisingly narrow range of wavelengths. The shortest wavelength blue light which is visible has a wavelength of $\lambda = 0.40$ microns (one micron is 10^{-6} meters). The longest wavelength red light which is visible has a wavelength of $\lambda = 0.76$ microns. However, there is nothing in Maxwell's analysis which suggested that this particular range of wavelengths is special. In principle, electromagnetic waves can have any wavelength. Maxwell concluded that visible light forms a small element of a vast spectrum of previously undiscovered types of electromagnetic radiation.

Since Maxwell's time, virtually all of the non-visible parts of the electromagnetic spectrum have been observed. Table gives a brief guide to the electromagnetic spectrum. Electromagnetic waves are of particular importance because they are our only source of information regarding the Universe around us. Radio waves and microwaves (which are comparatively hard to scatter) have provided much of our knowledge about the centre of the Galaxy. This is completely unobservable in visible light, which is strongly scattered by interstellar gas and dust lying in the galactic plane. For the same reason, the spiral arms of the Galaxy can only be mapped out using radio waves. Infrared radiation is useful for detecting proto-stars which are not yet hot enough to emit visible radiation. Of course, visible radiation is still the mainstay of astronomy. Satellite based ultraviolet observations have yielded invaluable insights into the structure and distribution of distant galaxies. Finally, X-ray and γ-ray astronomy usually concentrates on exotic objects in the Galaxy such as pulsars and supernova remnants.

Table: The electromagnetic spectrum.

Radiation Type	Wavelength Range (*m*)
Gamma Rays	$<10^{-11}$
X-Rays	10^{-11} - 10^{-9}
Ultraviolet	10^{-9} - 10^{-7}
Visible	10^{-7} - 10^{-6}
Infrared	10^{-6} - 10^{-4}
Microwave	10^{-4} - 10^{-1}
TV-FM	10^{-1} - 10^{1}
Radio	$> 10^{1}$

EFFECT OF DIELECTRIC MATERIALS

It turns out that electromagnetic waves cannot propagate very far through a conducting medium before they are either absorbed or reflected. However, electromagnetic waves are able to propagate through transparent dielectric media without difficultly. The speed of electromagnetic waves propagating through a dielectric medium is given by

$$C' = \frac{c}{\sqrt{K}},$$

where K is the dielectric constant of the medium in question, and the velocity of light in a vacuum. Since K > 1for dielectric materials, we conclude that:

The velocity with which electromagnetic waves propagate through a dielectric medium is always less than the velocity with which they propagate through a vacuum.

ENERGY IN ELECTROMAGNETIC WAVES

The energy stored per unit volume in an electromagnetic wave is given by

$$w = \frac{\epsilon_0 E^2}{2} + \frac{B^2}{2\mu_0}$$

Since, $B = E/c$, for an electromagnetic wave, and $c = 1/\sqrt{\mu_0 \epsilon_0}$, the above expression yields

$$w = \frac{\epsilon_0 E^2}{2} + \frac{E^2}{2\mu_0 c^2} = \frac{\epsilon_0 E^2}{2} + \frac{\epsilon_0 E^2}{2}:$$

or

$$w = \epsilon_0 E^2.$$

It is clear, from the above, that half the energy in an electromagnetic wave is carried by the electric field, and the other half is carried by the magnetic field.

As an electromagnetic field propagates it transports energy. Let P be the power per unit area carried by an electromagnetic wave: *i.e.*, P is the energy transported per unit time across a unit cross-sectional area perpendicular to the direction in which the wave is traveling. Consider a plane electromagnetic wave propagating along the z-axis. The wave propagates a distance cdt along the z-axis in a time interval dt. If we consider a cross-sectional area A at right-angles to the -axis, then in a time dt the wave sweeps through a volume dv of space, where $dV = Acdt$. The amount of energy filling this volume is

$$dW = wdV = \epsilon_0 E^2 Acdt.$$

It follows, from the definition of P, that the power per unit area carried by the wave is given by

$$p = \frac{dW}{Adt} = \frac{\epsilon_0 E^2 Acdt}{Adt}$$

so that

$$P = \epsilon_0 E^2 c.$$

Since half the energy in an electromagnetic wave is carried by the electric field, and the other half is carried by the magnetic field, it is conventional to convert the above expression into a form involving both the electric and magnetic field strengths. Since, E = cB, we have

$$P = \epsilon_0 cE(cB) = \epsilon_0 c^2 EB = \frac{EB}{\mu_0}.$$

Thus,

$$P = \frac{EB}{\mu_0}.$$

Equation specifies the power per unit area transported by an electromagnetic wave at any given instant of time. The *peak* power is given by

$$P_0 = \frac{E_0 B_0}{\mu_0},$$

where E_0 and B_0 are the peak amplitudes of the oscillatory electric and magnetic fields, respectively. It is easily demonstrated that the *average* power per unit area transported by an electromagnetic wave is *half* the peak power, so that

$$S = \bar{P} = \frac{E_0 B_0}{2\mu_0} = \frac{\epsilon_0 cE_0^2}{2} = \frac{cB_0^2}{2\mu_0}.$$

The quantity S is conventionally termed the *intensity* of the wave.

ELECTROMAGNETIC ORIGINS OF LIGHT

a. Static electric force equation: Coulomb's Law
b. Static interaction between magnetic materials (compass needles, etc.)

   Neither static interaction requires a medium between the source and the receiver – the effect works just as well between charges or magnets in a vacuum.

c. Moving charges create magnetic fields: magnetism and electrical effects are related

d. Changes in velocity produce varying magnetic fields which produce electric fields, which produce ...

e. wave are generated that travel with velocity defined by parameters from static force equations: $k_e/k_m = c^2$

f. energy radiated increases very rapidly with acceleration; acceleration increases with frequency of velocity variations

g. fluctuating electric fields can interact with charges to transfer energy and momentum to them

The fields do not require a medium to propagate. This was very surprising 100 years ago, but it probably should not have been given that the static interactions between charges and magnets don't require a medium either. People were so convinced that a medium had to exist that they created one called the "ether" whose sole function was to support transmission of electromagnetic waves. The existence of the ether was finally disproved by a number of experiments; the most famous is the Michelson-Morley experiment in 1887.

Chapter 4

General Properties of Waves

There are two ways to view a wave: either as the disturbance in all of space at a single instant in time or as the disturbance as a function of time at a single point in space. In the first view, we would plot the magnitude of the wave disturbance at the given instant of time as a function of a distance coordinate and in the second view we would plot the magnitude of the wave disturbance at a given point in space as a function of time. Both of these pictures look the same except for the labels and units on the x-axis.

A transverse wave has its disturbance perpendicular to its direction of propagation. Waves on a string, electromagnetic waves and water waves are transverse.

A longitudinal wave has its disturbance along the long of its direction of propagation. Sounds waves are longitudinal – the pressure fluctuations of the wave are in the same direction as the wave travels.

Amplitude: Maximum displacement of wave quantity relative to the undisturbed, equilibrium position. (height of water wave, pressure of sound wave, maximum electric field, etc.)

The Intensity of a wave or the power radiated by a source are proportional to the *square* of the amplitude.

Frequency: Number of cycles per second of the wave quantity, measured in Hertz (cycles per second), kHz, MHz, etc.

The frequency is usually represented by the letter n (nu) and occasionally by the letter f. The observation of the frequency is made at a single point in space.

Wavelength: Distance between corresponding points on successive cycles (e.g., between wave crests). Measured in units of length (e.g., meters, nano-meters)The wavelength is usually represented by the letter l (lambda). A measurement of the wavelength is made by observing the wave in space at a single instant of time.

Period: Time between successive wave crests. Measured in units of time (e.g., seconds). The period is often represented by the letter T. The period is measured by observing the wave displacement at a single point in space.

Phase:The time at which a wave passes through a specified displacement at a specified point in space. A common specification is the time at which the wave disturbance is exactly zero and moving in the positive direction. Although there are some applications for the absolute specification of phase, it is more common to speak of the phase *difference* between two waves.Although this difference can have any value in principle, the two situations we will usually encounter in this course are:

1. Two waves that are exactly in phase – that is the maxima and minima of the two waves occur at exactly the same instants of time and
2. Two waves that are exactly out of phase – that is the maximum of one wave occurs at the same instant as the minimum of the other. Phase differences can be measured in units of time, but are more commonly measured in fractions of a cycle. Two waves that are in phase have a phase difference of exactly 0, while two waves that are exactly out of phase have a phase difference of exactly one-half of a cycle.

Velocity of propagation: Speed of wave in space. Measured in units of speed (e.g., meters/second, km/s, etc.)The velocity of propagation is usually a characteristic of the medium and the type of wave – it doesnot depend on the parameters of the source or the detector.The velocity of light in vacuum is usually represented by the letter c. The velocity of light in some medium other than a vacuum (such as air) or the velocity of some other kind of wave is usually represented by v. However, we will sometimes use c in this case as well, to avoid confusing between the velocity v and the frequency n.

Polarization:The direction of the wave disturbance in space. An electromagnetic wave that is "vertically polarized" has its electric field vector oscillating vertically (from straight up to straight down). A wave that was "horizontally polarized" would have its electric field vector oscillating horizontally (from right to left). Intermediate states of polarization between purely vertical and purely horizontal are also possible. These cases are often identified by the angle of the wave quantity with respect to the vertical axis.

The following relationships are true for all waves:

$n\lambda = c$ (product of wavelength and frequency equals velocity of propagation)

$n = 1/T$

$T= 1/n$ (frequency and period are reciprocals of each other)

The safest method is to convert all velocity and distance quantities to meters/second or meters and all time-based quantities to seconds or Hz, removing any metric prefixes that may be present. The velocity of light can be taken to be 3×10^8 m/s in all calculations. You should remember that this number is not exactly correct, but it is close enough for most purposes.

THE SOURCES OF ELECTROMAGNETIC RADIATION

General principle: The size of a transmitting antenna is about the same as the size of the wave it produces For electromagnetic waves, it is common to use an antenna that is one-quarter of the wavelength of the wave to be generated.(By comparison, sources of sound waves often generate many wavelengths simultaneously. They usually have a length that is one-half or one-quarter of the longest wavelength (lowest frequency) that can be generated.)

a. Moving charges in wires and bulk materials are usually used at low frequencies physical oscillation of the charges matches frequency of the radiated energy

typical wavelengths

AM radio	(1 MHz)	300 m
FM radio	(100 MHz)	3 m
Cell telephone	(900 MHz)	0.33 m
Microwave oven	(3 GHz)	0.1 m

Large-scale movement of charges in wires and bulk materials becomes impractical at higher frequencies

b. The black body radiation function electromagnetic energy generated by random motions of charges in a material radiation is not at one single frequency details of frequency spectrum are not obvious and not trivial to calculate – requires introducing quantum phenomena Total power increases very rapidly with temperature (increases as T^4), where T is measured in Kelvin (degrees centigrade+273).

Peak of radiation function moves to higher frequency/ shorter wavelengths as temperature increases. The power radiated at visible wavelengths begins to become appreciable when the temperature reaches about 600 K (350° C), and objects at that temperature begin to appear red to the eye.

c. Atomic and molecular transmitters and receivers resonance behaviour due to details of atomic and molecular structure – atoms and molecules absorb and emit energy only at specific frequencies/ wavelengths.

DETECTING ELECTROMAGNETIC RADIATION

General principle: For radio signals, the size of the receiving antenna is about the same as the size of the wavelength it is designed to receive.

For electromagnetic waves used for communications (radio, TV, cell phones, etc.), it is common to use an antenna that is one-quarter of the wavelength of the wave to be generated. Unlike a transmitter, the size of the antenna is often driven by other considerations – the physical size of a cell phone, for example.

In addition to tuning the receiver based on the length of the antenna, there are usually additional frequency-selective elements in a receiver. The radio frequency portion of the spectrum extends from very low frequencies on the order of kHz to frequencies of 20 GHz (or even higher in some applications). The corresponding wavelengths range from hundreds of meters for the low frequencies to several centimeters for the highest frequencies so that it is feasible to construct antennas that are comparable in size to the wavelength of the desired signal. The wavelengths of optical signals are too short to design antennas based on these considerations. Optical detectors tend to be based on the interaction between the electromagnetic radiation and atoms or molecules, which often respond only to specific wavelengths or to a specific range of wavelengths. The details of the detection mechanism depend on the detailed properties of the material, and understanding these mechanisms requires complex calculations using quantum mechanics.

Since a black body absorbs all of the energy that is incident on it and uses the energy to raise its internal temperature, the temperature of a black body can be used as a relatively crude indicator of the size of a received signal. By definition (a black body absorbs *everything* equally well), this sort of detector cannot provide any information on the frequency of the incident radiation. A very sensitive detector based on this principle is called a bolometer, which is basically a black body combined with a very sensitive thermometer.

Another common detection method uses the photoelectric effect, in which an incident light wave transfer enough energy to an electron in the receiving material to cause the electron to escape from the material. The energy with which the electron is ejected from the material is related to the wavelength of the incident radiation. This electron can be captured by a nearby electrode, and it can be detected in this way. Variations on this idea are used in television cameras, to turn street lights on at dusk, etc. As with other interactions between light and matter, a detailed understanding of the photoelectric effect requires quantum mechanics.

In addition to discrimination based on the frequency or wavelength of a signal, many receivers include complex pattern-matching algorithms. These algorithms are gradually growing more complex as the spectrum becomes more crowded and more applications share the same frequency allocation. Pattern-matching algorithms are common in cell phones, pagers, and optical processing. They are especially important in real-world optical applications, such as distinguishing between red T-shirts and red traffic lights.

Chapter 5

Geometrical Optics

INTRODUCTION

Optics deals with the propagation of light through transparent media, and its interaction with mirrors, lenses, slits, *etc.* Optical effects can be divided into two broad classes. Firstly, those which can be explained without reference to the fact that light is fundamentally a wave phenomenon, and, secondly, those which can only be explained on the basis that light is a wave phenomenon. Let us, for the moment, consider the former class of effects. It might seem somewhat surprising that any optical effects at all can be accounted for without reference to waves. light really is a wave phenomenon. It turns out, however, that wave effects are only crucially important when the wavelength of the wave is either comparable to, or much larger than, the size of the objects with which it interacts.When the wavelength of the wave becomes much smaller than the size of the objects with which it interacts then the interactions can be accounted for in a very simple geometric manner, as explained in this section. Since the wavelength of visible light is only of order a micron, it is very easy to find situations in which its wavelength is very much smaller than the size of the objects with which it interacts. Thus, "wave-less" optics, which is usually called *geometric optics*, has a very wide range of applications.

In geometric optics, light is treated as a set of *rays*, emanating from a source, which propagate through transparent media according to a set of *three* simple laws. The first law is the *law of rectilinear propagation,* which states that light rays propagating through a homogeneous transparent medium do so in straight-lines. The second law is the *law of reflection,* which governs the interaction of light rays with conducting surfaces (*e.g.,* metallic mirrors). The third law is the *law of refraction,* which governs the behaviour of light rays as they traverse a sharp boundary between two different transparent media (*e.g.,* air and glass).

HISTORY OF GEOMETRIC OPTICS

Let us first consider the law of *rectilinear propagation.* The earliest surviving optical treatise, Euclid's *Catoptrics,* recognized that light travels in straight-lines in homogeneous media. However, following the teachings of Plato, Euclid (and all other ancient Greeks) thought that light rays emanate from the eye, and intercept external objects, which are thereby "seen" by the observer. The ancient Greeks also thought that the speed with which light rays emerge from the eye is very high, if not infinite. After all, they argued, an observer with his eyes closed can open them and immediately see the distant stars.

Hero of Alexandria, in his *Catoptrics* (first century BC), also maintained that light travels with infinite speed. His argument was by analogy with the free fall of objects. If we throw an object horizontally with a relatively small velocity then it manifestly does not move in a straight-line. However, if we throw an object horizontally with a relatively large velocity then it appears to move in a straight-line to begin with, but eventually deviates from this path. The larger the velocity with which the object is thrown, the longer the initial period of apparent rectilinear motion. Hero reasoned that if an object were thrown with an infinite velocity then it would move in a straight-line forever. Thus, light, which travels in a straight-line, must move with an infinite velocity. The erroneous idea that light travels with an *infinite* velocity persisted until 1676, when the Danish astronomer Olaf Römer demonstrated that light must have a *finite* velocity, using his timings of the successive eclipses of the satellites of Jupiter, as they passed into the shadow of the planet.

The first person to realise that light actually travels from the object seen to the eye was the Arab philosopher "Alhazan" (whose real name was Abu'ali al-hasan ibn al-haytham), who published a book on optics in about 1000 AD.

The law of *reflection* was correctly formulated in Euclid's book. Hero of Alexandria demonstrated that, by adopting the rule that light rays always travel between two points by the *shortest path* (or, more rigorously, the extremal path), it is possible to derive the law of reflection using geometry.

The law of *refraction* was studied experimentally by Claudius Ptolemy (100-170AD), and is reported in Book V of his *Catoptrics*. Ptolemy formulated a very inaccurate version of the law of refraction, which only works when the light rays are almost normally incident on the interface in question. Despite its obvious inaccuracy, Ptolemy's theory of refraction persisted for nearly 1500 years. The true law of refraction was discovered empirically by the Dutch mathematician Willebrord Snell in 1621. However, the French philosopher René Descartes was the first to publish, in his *La Dioptrique* (1637), the now familiar formulation of the law of refraction in terms of sines. Although there was much controversy at the time regarding plagiarism, Descartes was apparently unaware of Snell's work. Thus, in English speaking countries the law of refraction is called "Snell's law", but in French speaking countries it is called "Descartes' law".

In 1658, the French mathematician Pierre de Fermat demonstrated that all three of the laws of geometric optics can be accounted for on the assumption that light always travels between two points on the path which takes the *least time* (or, more rigorously, the extremal time). Fermat's ideas were an extension of those of Hero of Alexandria. Fermat's (correct) derivation of the law of refraction depended crucially on his (correct) assumption that light travels *more slowly* in dense media than it does in air. Unfortunately, many famous scientists, including Newton, maintained that light travels *faster* in dense media than it does in air. This erroneous idea held up progress in optics for over one hundred years, and was not conclusively disproved until the mid-nineteenth century. Incidentally, Fermat's principle of least time can only be justified using wave theory.

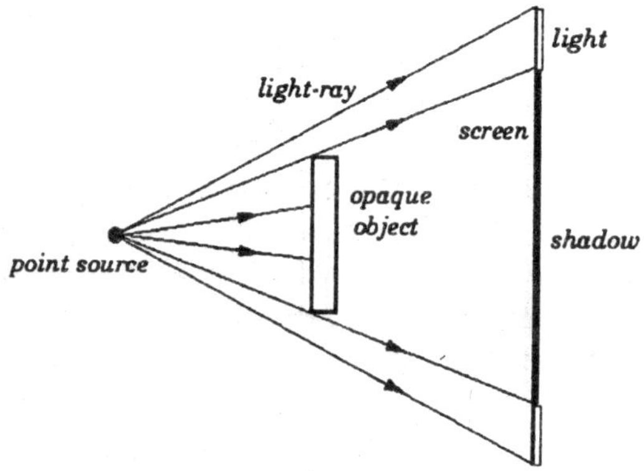

Fig. An opaque object illuminated by a point light source.

LAW OF GEOMETRIC PROPAGATION

According to geometric optics, an opaque object illuminated by a point source of light casts a sharp shadow whose dimensions can be calculated using geometry. The method of calculation is very straightforward. The source emits *light-rays* uniformly in all directions. These rays can be represented as straight lines radiating from the source. The light-rays propagate away from the source until they encounter an opaque object, at which point they stop. This is illustrated in Fig.

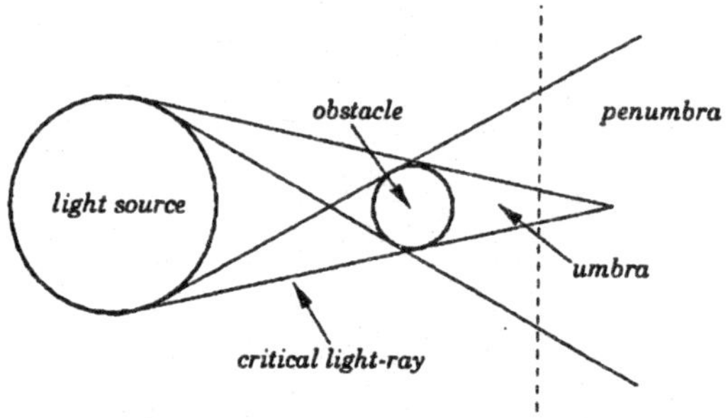

Fig. An opaque object illuminated by an extended light source.

For an extended light source, each element of the source emits light-rays, just like a point source. Rays emanating from different elements of the source are assumed not to interfere with one another. Figure shows how the shadow cast by an opaque sphere illuminated by a spherical light source is calculated using a small number of critical light-rays. The shadow consists of a perfectly black disk called the *umbra,* surrounded by a ring of gradually diminishing darkness called the *penumbra.* In the umbra, *all* of the light-rays emitted by the source are blocked by the opaque sphere, whereas in the penumbra only *some* of the rays emitted by the source are blocked by the sphere. As was well-known to the ancient Greeks, if the light-source represents the Sun, and the opaque sphere the Moon, then at a point on the Earth's surface which is situated inside the umbra the Sun is totally eclipsed, whereas at a point on the Earth's surface which is situated in the penumbra the Sun is only partially eclipsed.

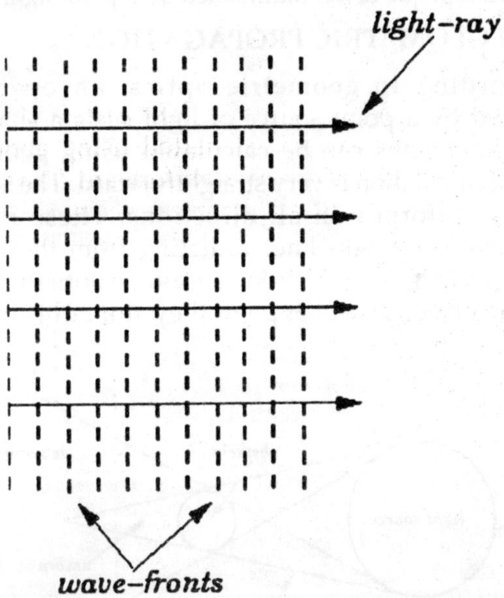

Fig. Relationship between wave-fronts and light-rays.

In the wave picture of light, a *wave-front* is defined as a surface joining all adjacent points on a wave that have the same phase (*e.g.*, all maxima, or minima, of the electric field). A light-ray is simply a line which runs perpendicular to the wave-fronts at all points along the path of the wave. This is illustrated in Fig. Thus, the law of rectilinear propagation of light-rays also specifies how wave-fronts propagate through homogeneous media. Of course, this law is only valid in the limit where the wavelength of the wave is much smaller than the dimensions of any obstacles which it encounters.

LAW OF REFLECTION

The law of reflection governs the reflection of light-rays off smooth conducting surfaces, such as polished metal or metal-coated glass mirrors.

Consider a light-ray incident on a plane mirror, as shown in Fig. The law of reflection states that the incident ray, the reflected ray, and the normal to the surface of the mirror all lie in the *same plane*. Furthermore, the angle of reflection is *equal* to the angle of incidence. Both angles are measured with respect to the normal to the mirror.

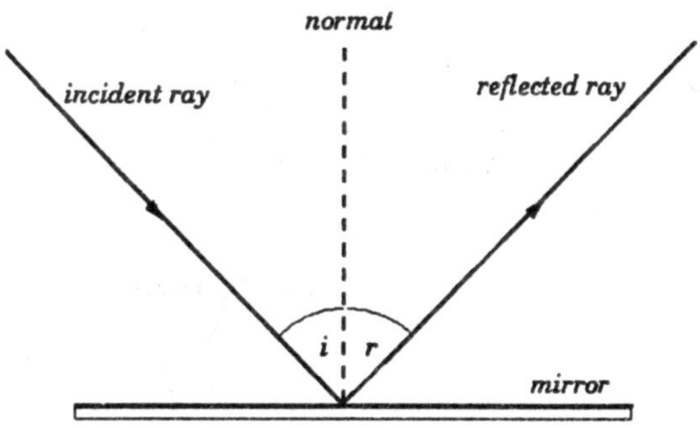

Fig. The law of reflection

The law of reflection also holds for non-plane mirrors, provided that the normal at any point on the mirror is understood to be the outward pointing normal to the local tangent plane of the mirror at that point. For rough surfaces, the law of reflection remains valid. It predicts that rays incident at slightly different points on the surface are reflected in completely different directions, because the normal to a rough surface varies in direction very strongly from point to point on the surface. This type of reflection is called *diffuse reflection*, and is what enables us to see non-shiny objects.

LAW OF REFRACTION

The law of refraction, which is generally known as *Snell's law*, governs the behaviour of light-rays as they propagate across a sharp interface between two transparent dielectric media.

Consider a light-ray incident on a plane interface between two transparent dielectric media, labelled 1 and 2, as shown in Fig. The law of refraction states that the incident ray, the refracted ray, and the normal to the interface, all lie in the *same plane*. Furthermore,

$$n_1 \sin\theta_1 = n_2 \sin\theta_2$$

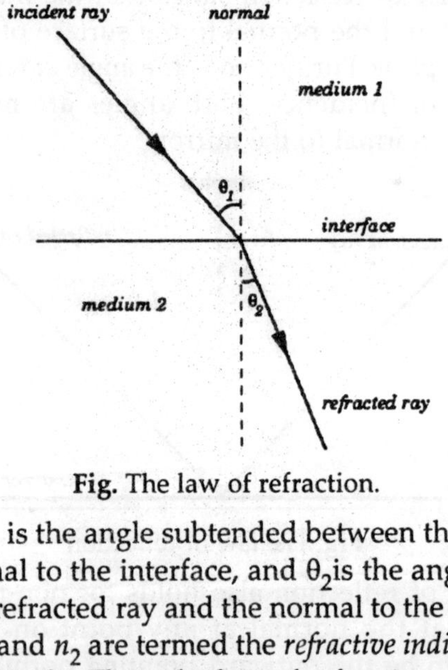

Fig. The law of refraction.

where θ_1 is the angle subtended between the incident ray and the normal to the interface, and $\theta_2$is the angle subtended between the refracted ray and the normal to the interface. The quantities n_1 and n_2 are termed the *refractive indices* of media 1 and 2, respectively. Thus, the law of refraction predicts that a light-ray always deviates more towards the normal in the optically denser medium: *i.e.*, the medium with the higher refractive index. Note that n2 > n1in the figure. The law of refraction also holds for non-planar interfaces, provided that the normal to the interface at any given point is understood to be the normal to the local tangent plane of the interface at that point.

By definition, the refractive index n of a dielectric medium of dielectric constant K is given by

$$n = \sqrt{K}$$

Table shows the refractive indices of some common materials (for yellow light of wavelength λ = 589 nm).

Table: Refractive indices of some common materials at λ = 589 nm.

Material	n
Air (STP)	1.00029
Water	1.33
Ice	1.31
Glass:	
Light flint	1.58
Heavy flint	1.65
Heaviest flint	1.89
Diamond	2.42

The law of refraction follows directly from the fact that the speed v with which light propagates through a dielectric medium is *inversely proportional* to the refractive index of the medium In fact,

$$v = \frac{c}{n},$$

where c is the speed of light in a vacuum. Consider two parallel light-rays, a and b, incident at an angle θ_1 with respect to the normal to the interface between two dielectric media, 1 and 2. Let the refractive indices of the two media be n_1 and n_2 respectively, with. It is clear from Fig. that ray b must move from point to B point Q, in medium 1, in the same time interval,, in which ray moves between points and Δt, in medium 2. Now, the speed of light in medium 1 is $v_1 = c/n_1$, whereas the speed of light in medium 2 is $v_2 = c/n_2$. It follows that the length BQ is given by $v_1\Delta t$, whereas the length AP is given by. $v_2\Delta t$By trigonometry,

$$\sin\theta_1 = \frac{BQ}{AQ} = \frac{v_1\Delta t}{AQ}$$

and

$$\sin\theta 2 = \frac{AP}{AQ} = \frac{v_2\Delta t}{AQ}.$$

Hence,

$$\frac{\sin\theta_1}{\sin\theta_2} = \frac{v_1}{v_2} = \frac{n_2}{n_1},$$

which can be rearranged to give Snell's law. Note that the lines AB and PQ represent wave-fronts in media 1 and 2, respectively, and, therefore, cross rays a and b at right-angles.

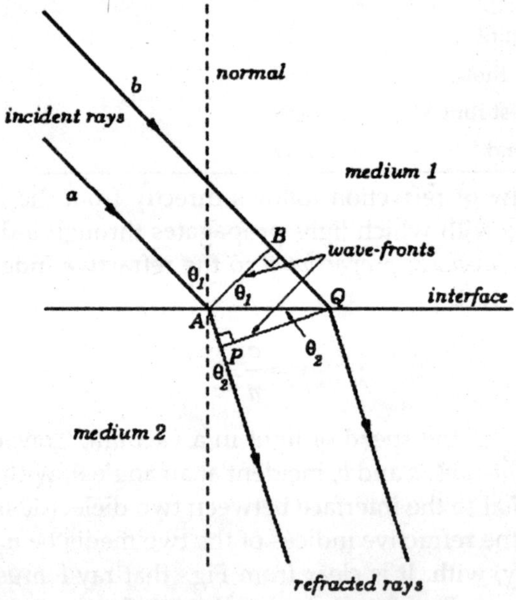

Fig. Derivation of Snell's law.

When light passes from one dielectric medium to another its velocity changes, but its frequency f remains *unchanged.*

Since, $v = f\lambda$ for all waves, where λ is the wavelength, it follows that the wavelength of light must also change as it crosses an interface between two different media. Suppose that light propagates from medium 1 to medium 2. Let n_1 and be the refractive indices of the two media, respectively. The ratio of the wave-lengths in the two media is given by

$$\frac{\lambda_2}{\lambda_1} = \frac{v_2/f}{v_1/f} = \frac{v_2}{v_1} = \frac{n_1}{n_2}.$$

Thus, as light moves from air to glass its wavelength *decreases.*

TOTAL INTERNAL REFLECTION

An interesting effect known as *total internal reflection* can occur when light attempts to move from a medium having a given refractive index to a medium having a *lower* refractive index. Suppose that light crosses an interface from medium 1 to medium 2, where $n_2 < n_1$. According to Snell's law,

$$\sin\theta_2 = \frac{n_1}{n_2}\sin\theta_1.$$

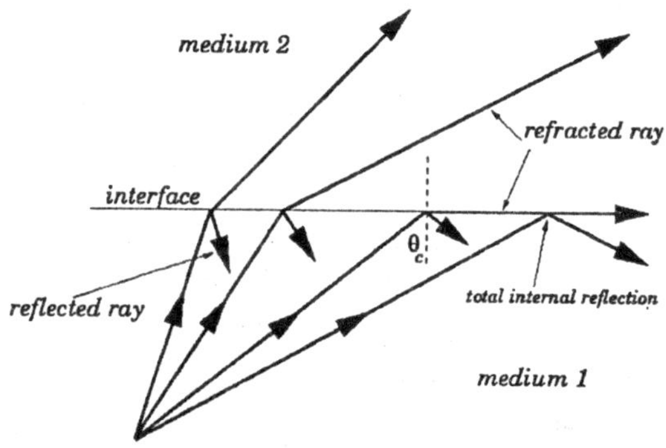

Fig. Total internal reflection.

Since $n_1/n_s > 1$, it follows that $\theta_2 > \theta_1$. For relatively small angles of incidence, part of the light is refracted into the less optically dense medium, and part is reflected (there is always some reflection at an interface). When the angle of incidence θ_1 is such that the angle of refraction $\theta_2 = 90°$, the refracted ray runs along the interface between the two media. This particular angle of incidence is called the *critical angle,* θ_c. For $\theta_1 > \theta_c$, there is *no* refracted ray. Instead, all of the light incident on the interface is reflected—see Fig. This effect is called *total internal reflection,* and occurs whenever the angle of incidence exceeds the critical angle. Now when $\theta_1 = \theta_c$, we have $\theta_2 = 90°$, and so $\sin\theta_2 = 1$. It follows from Equation. that

$$\sin\theta_c = \frac{n_2}{n_1}.$$

Consider a fish (or a diver) swimming in a clear pond. As Fig. makes clear, if the fish looks upwards it sees the sky, but if it looks at too large an angle to the vertical it sees the bottom of the pond reflected on the surface of the water. The critical angle to the vertical at which the fish first sees the reflection of the bottom of the pond is, of course, equal to the critical angle for total internal reflection at an air-water interface. From Equation. (this critical angle is given by

$$\theta_c = \sin^{-1}(1.00/1.33) = 48.8°,$$

since the refractive index of air is approximately unity, and the refractive index of water is 1.33.

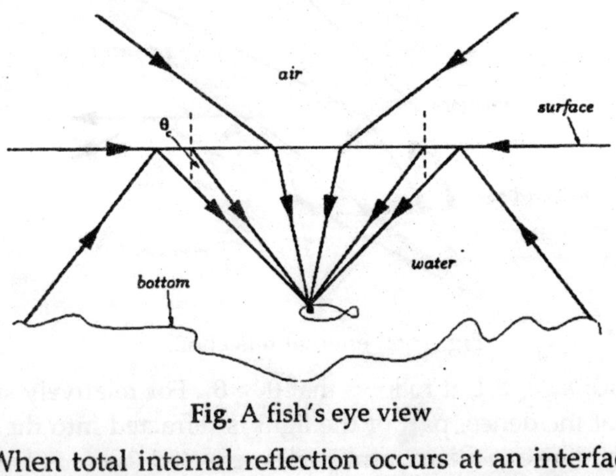

Fig. A fish's eye view

When total internal reflection occurs at an interface the interface in question acts as a *perfect reflector*. This allows crown glass prisms to be used, in place of mirrors, to reflect light in binoculars. This is illustrated in Fig. The angles of incidence on the sides of the prism are all, which is greater than the critical angle for crown glass (at an air-glass interface).

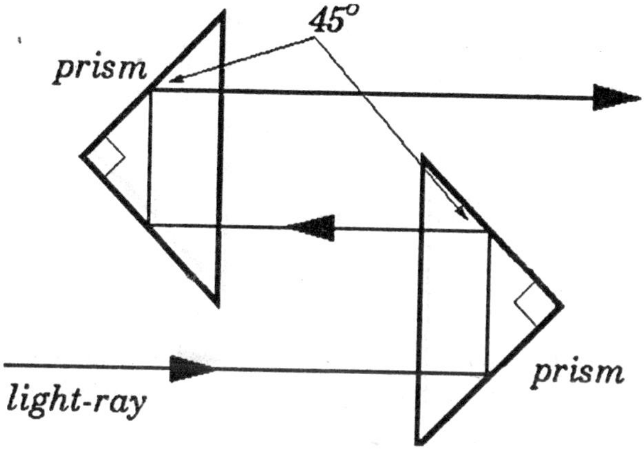

Fig. Arrangement of prisms used in binoculars.

Diamonds, for which, have a critical angle which is only. The facets on a diamond are cut in such a manner that much of the incident light on the diamond is reflected many times by successive total internal reflections before it escapes. This effect gives rise to the characteristic sparkling of cut diamonds.

Total internal reflection enables light to be transmitted inside thin glass fibers. The light is internally reflected off the sides of the fiber, and, therefore, follows the path of the fiber. Light can actually be transmitted around corners using a glass fiber, provided that the bends in the fiber are not too sharp, so that the light always strikes the sides of the fiber at angles greater than the critical angle. The whole field of *fiber optics,* with its many useful applications, is based on this effect.

DISPERSION

When a wave is refracted into a dielectric medium whose refractive index *varies* with wavelength then the angle of refraction also varies with wavelength. If the incident wave is not monochromatic, but is, instead, composed of a mixture of waves of different wavelengths, then each component wave is refracted through a *different* angle. This phenomenon is called *dispersion.*

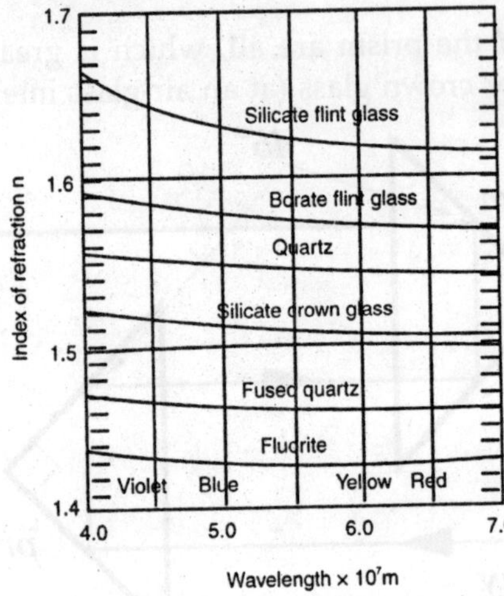

Fig. Refractive indices of some common materials as functions of wavelength.

Figure shows the refractive indices of some common materials as functions of wavelength in the visible range. It can be seen that the refractive index always *decreases* with increasing wavelength in the visible range. In other words, violet light is always refracted *more strongly* than red light.

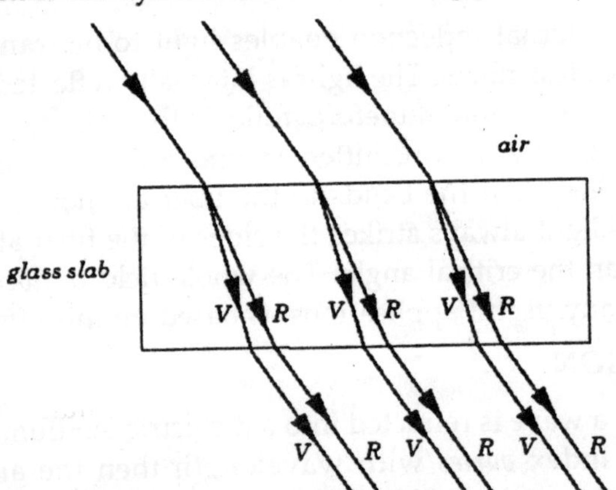

Fig. Dispersion of light by a parallel-sided glass slab.

Suppose that a parallel-sided glass slab is placed in a beam of white light. Dispersion takes place inside the slab, but, since the rays which emerge from the slab all run *parallel* to one another, the dispersed colours recombine to form white light again, and no dispersion is observed except at the very edges of the beam. This is illustrated in Fig. It follows that the dispersion of white light through a parallel-sided glass slab is not generally a noticeable effect.

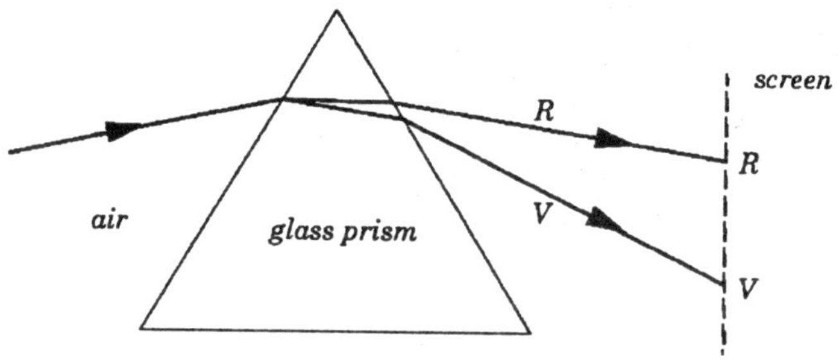

Fig. Dispersion of light by a glass prism.

Suppose that a glass prism is placed in a beam of white light. Dispersion takes place inside the prism, and, since the emerging rays are *not parallel* for different colours, the dispersion is clearly noticeable, especially if the emerging rays are projected onto a screen which is placed a long way from the prism. This is illustrated in Fig. It is clear that a glass prism is far more effective at separating white light into its component colours than a parallel-sided glass slab (which explains why prisms are generally employed to perform this task).

RAINBOWS

The most well-known, naturally occurring phenomenon which involves the dispersion of light is a *rainbow*. A rainbow is an *arc* of light, with an angular radius of 42°, centred on a direction which is *opposite* to that of the Sun in the sky (*i.e.*, it is centred on the direction of propagation of the Sun's rays. Thus, if the Sun is low in the sky (*i.e.*, close to the horizon) we see almost a full semi-circle. If the Sun is higher in the sky we see a smaller arc, and if the Sun is more than 42° above the horizon then there is no rainbow (for viewers on the Earth's surface). Observers on a hill may see parts of the rainbow below the horizontal: *i.e.,* an arc greater than a semi-circle. Passengers on an airplane can sometimes see a full circle.

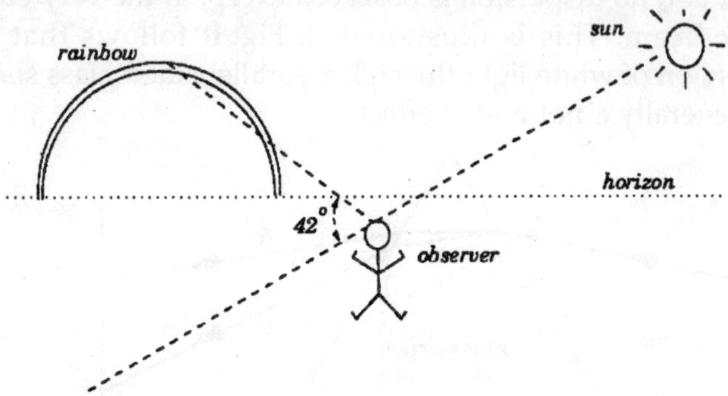

Fig. A rainbow

The colours of a rainbow vary smoothly from red on the outside of the arc to violet on the inside. A rainbow has a diffuse inner edge, and a sharp outer edge. Sometimes a *secondary arc* is observed. This is fainter and larger (with an angular radius of 50°) than the primary arc, and the order of the colours is reversed (*i.e.,* red is on the inside, and violet on the outside). The secondary arc has a diffuse outer edge, and a sharp inner edge. The sky between the two arcs sometimes appears to be less bright than the sky elsewhere. This region is called *Alexander's dark band,* in honour of Alexander of Aphrodisias who described it some 1800 years ago.

Rainbows have been studied since ancient times. Aristotle wrote extensively on rainbows in his *De Meteorologica,* and even speculated that a rainbow is caused by the reflection of sunlight from the drops of water in a cloud.

The first scientific study of rainbows was performed by Theodoric, professor of theology at Freiburg, in the fourteenth century. He studied the path of a light-ray through a spherical globe of water in his laboratory, and suggested that the globe be thought of as a model of a single falling raindrop. A ray, from the Sun, entering the drop, is refracted at the air-water interface, undergoes internal reflection from the inside surface of the drop, and then leaves the drop in a backward direction, after being again refracted at the surface. Thus, looking away from the Sun, towards a cloud of raindrops, one sees an enhancement of light due to these rays. Theodoric did not explain why this enhancement is concentrated at a particular angle from the direction of the Sun's rays, or why the light is split into different colours.

The first person to give a full explanation of how a rainbow is formed was René Descartes. He showed mathematically that if one traces the path through a spherical raindrop of parallel light-rays entering the drop at different points on its surface, each emerges in a different direction, but there is a concentration of emerging rays at an angle of 42° from the reverse direction to the incident rays, in exact agreement with the observed angular size of rainbows. Furthermore, since some colours are refracted more than others in a raindrop, the "rainbow angle" is slightly different for each colour, so a raindrop disperses the Sun's light into a set of nearly overlapping coloured arcs.

Figure illustrated Descartes' theory in more detail. It shows parallel light-rays entering a spherical raindrop. Only rays entering the upper half contribute to the rainbow effect. Let us follow the rays, one by one, from the top down to the middle of the drop. We observe the following pattern. Rays which enter near the top of the drop emerge going in almost the reverse direction, but a few degrees below the horizontal. Rays entering a little further below the top emerge at a greater angle below the horizontal. Eventually, we reach a critical ray, called the *rainbow ray,* which emerges in an angle 42° below the horizontal. Rays entering the drop lower than the rainbow ray emerge at an angle less than 42°. Thus, the rainbow ray is the one which deviates *most* from the reverse direction to the incident rays. This variation, with 42° being the maximum angle of deviation from the reverse direction, leads to a bunching of rays at that angle, and hence, to an unusually bright arc of reflected light centred around from the reverse direction. The arc has a sharp outer edge, since reflected light *cannot* deviate by more than 42° from the reverse direction, and a diffuse inner edge, since light *can* deviate by less than 42° from the reverse direction: 42° is just the *most likely* angle of deviation. Finally, since the rainbow angle varies slightly with wavelength (because the refractive index of water varies slightly with wavelength), the arcs corresponding to each colour appear at slightly different angles relative to the reverse direction to the incident rays. We expect violet light to be refracted more strongly than red light in a raindrop. It is, therefore, clear, from Fig.that the red arc deviates slightly more from the reverse direction to the incident rays than the violet arc. In other words, violet is concentrated on the inside of the rainbow, and red is concentrated on the outside.

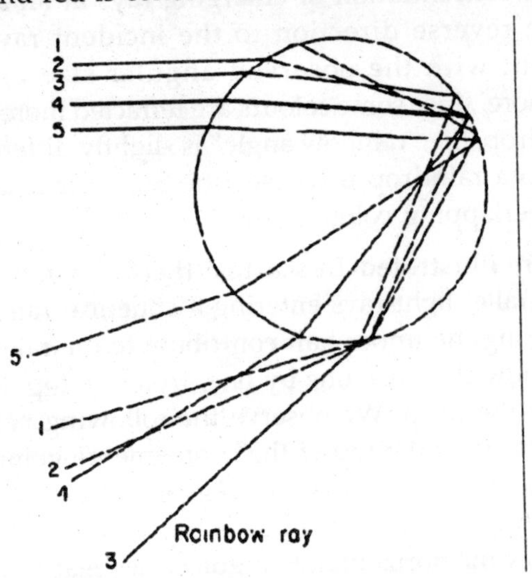

Fig. Descarte's theory of the rainbow.

Descartes was also able to show that light-rays which are internally reflected *twice* inside a raindrop emerge concentrated at an angle of from the reverse direction to the incident rays. Of course, this angle corresponds exactly to the angular size of the secondary rainbow sometimes seen outside the first. This rainbow is naturally less intense than the primary rainbow, since a light-ray loses some of its intensity at each reflection or refraction event. Note that represents the angle of *maximum* deviation of doubly reflected light from the reverse direction (*i.e.*, doubly reflected light can deviate by more than this angle, but not by less). Thus, we expect the secondary rainbow to have a diffuse outer edge, and a sharp inner edge. We also expect doubly reflected violet light to be refracted more strongly in a raindrop than doubly reflected red light. It follows, from Fig, that the red secondary arc deviates slightly less from the reverse direction to the incident rays than the violet secondary arc. In other words, red is concentrated on the inside of the secondary rainbow, and violet on the outside. Since no reflected light emerges between the primary and secondary rainbows (*i.e.*, in the angular range to, relative to the reverse direction), we naturally expect this region of the sky to look slightly less bright than the other surrounding regions of the sky, which explains Alexander's dark band.

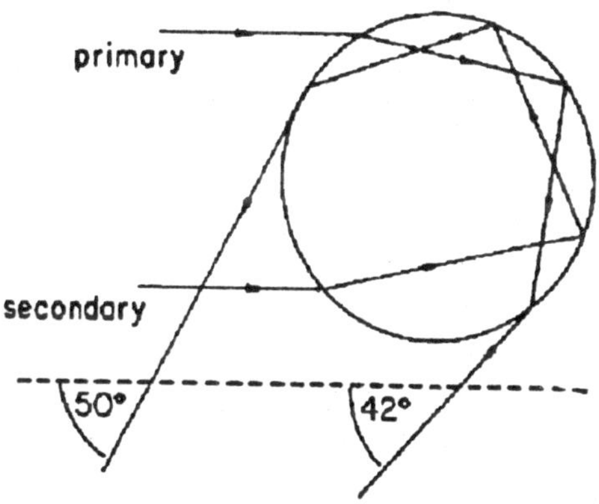

Fig. Rainbow rays for the primary and secondary arcs of a rainbow.

PARAXIAL OPTICS

SPHERICAL MIRRORS

A spherical mirror is a mirror which has the shape of a piece cut out of a spherical surface. There are two types of spherical mirrors: *concave*, and *convex*. These are illustrated in Fig.The most commonly occurring examples of concave mirrors are shaving mirrors and makeup mirrors. As is well-known, these types of mirrors magnify objects placed close to them.

The most commonly occurring examples of convex mirrors are the passenger-side wing mirrors of cars. These type of mirrors have wider fields of view than equivalent flat mirrors, but objects which appear in them generally look smaller (and, therefore, farther away) than they actually are.

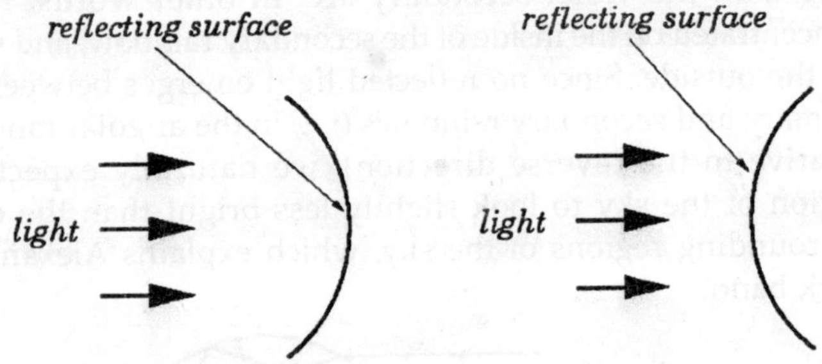

Fig. A concave (left) and a convex (right) mirror

Let us now introduce a few key concepts which are needed to study image formation by a concave spherical mirror. As illustrated in Fig. the normal to the centre of the mirror is called the *principal axis*. The mirror is assumed to be *rotationally symmetric* about this axis. Hence, we can represent a three-dimensional mirror in a two-dimensional diagram, without loss of generality. The point V at which the principal axis touches the surface of the mirror is called the *vertex*. The point C, on the principal axis, which is equidistant from all points on the reflecting surface of the mirror is called the *centre of curvature*. The distance along the principal axis from point C to point is V called the *radius of curvature* of the mirror, and is denoted. It is found experimentally that rays striking a concave mirror parallel to its principal axis, and not too far away from this axis, are reflected by the mirror such that they all pass through the same point F on the principal axis. This point, which is lies between the centre of curvature and the vertex, is called the *focal point*, or *focus*, of the mirror. The distance along the principal axis from the focus to the vertex is called the *focal length* of the mirror, and is denoted f.

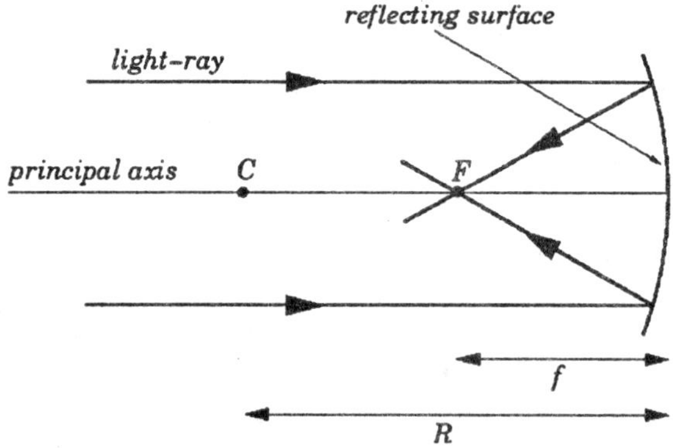

Fig. Image formation by a concave mirror.

In our study of concave mirrors, we are going to assume that all light-rays which strike a mirror parallel to its principal axis (*e.g.*, all rays emanating from a distant object) are brought to a focus at the same point *F*. Of course, as mentioned above, this is only an approximation. It turns out that as rays from a distant object depart further from the principal axis of a concave mirror they are brought to a focus ever closer to the mirror, as shown in Fig. This lack of perfect focusing of a spherical mirror is called *spherical aberration*. The approximation in which we neglect spherical aberration is called the *paraxial approximation*.Likewise, the study of image formation under this approximation is known as *paraxial optics*. This field of optics was first investigated systematically by the famous German mathematician Karl Friedrich Gauss in 1841.

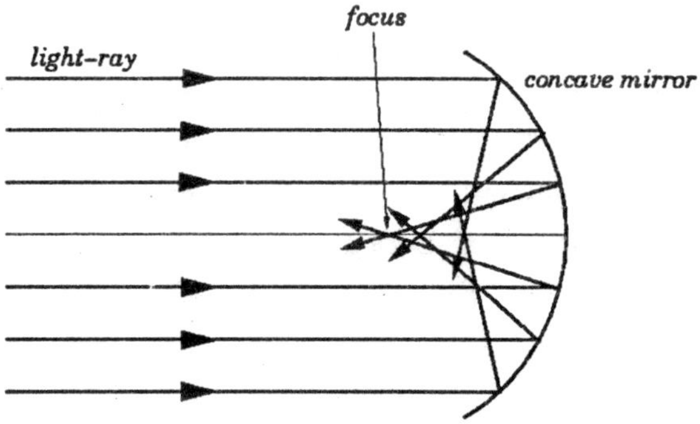

Fig. Spherical aberration in a concave mirror.

Can be demonstrated, by geometry, that the only type of mirror which does not suffer from spherical aberration is a parabolic mirror (i.e., a mirror whose reflecting surface is the surface of revolution of a parabola). Thus, a ray traveling parallel to the principal axis of a parabolic mirror is brought to a focus at the same point *F*, no matter how far the ray is from the axis. Since the path of a light-ray is completely reversible, it follows that a light source placed at the focus *F* of a parabolic mirror yields a perfectly parallel beam of light, after the light has reflected off the surface of the mirror. Parabolic mirrors are more difficult, and, therefore, more expensive, to make than spherical mirrors. Thus, parabolic mirrors are only used in situations where the spherical aberration of a conventional spherical mirror would be a serious problem. The receiving dishes of radio telescopes are generally parabolic. They reflect the incoming radio waves from (very) distant astronomical sources, and bring them to a focus at a single point, where a detector is placed. In this case, since the sources are extremely faint, it is imperative to avoid the signal losses which would be associated with spherical aberration. A car headlight consists of a light-bulb placed at the focus of a parabolic reflector. The use of a parabolic reflector enables the headlight to cast a very straight beam of light ahead of the car. The beam would be nowhere near as well-focused were a spherical reflector used instead.

IMAGE FORMATION BY CONCAVE MIRRORS

There are two alternative methods of locating the image formed by a concave mirror. The first is purely graphical, and the second uses simple algebraic analysis.

The graphical method of locating the image produced by a concave mirror consists of drawing light-rays emanating from key points on the object, and finding where these rays are brought to a focus by the mirror. This task can be accomplished using just *four* simple rules:

1. An incident ray which is parallel to the principal axis is reflected through the focus of the mirror.

2. An incident ray which passes through the focus of the mirror is reflected parallel to the principal axis.
3. An incident ray which passes through the centre of curvature of the mirror is reflected back along its own path (since it is normally incident on the mirror).
4. An incident ray which strikes the mirror at its vertex is reflected such that its angle of incidence with respect to the principal axis is equal to its angle of reflection.

The validity of these rules in the paraxial approximation is fairly self-evident.

Consider an object ST which is placed a distance p from a concave spherical mirror, as shown in Fig.For the sake of definiteness, let us suppose that the object distance p is greater than the focal length f of the mirror. Each point on the object is assumed to radiate light-rays in all directions. Consider four light-rays emanating from the tip T of the object which strike the mirror, as shown in the figure. The reflected rays are constructed using rules 1-4 above, and the rays are labelled accordingly. It can be seen that the reflected rays all come together at some point T'. Thus, T' is the image of T (i.e., if we were to place a small projection screen at T' then we would see an image of the tip on the screen). As is easily demonstrated, rays emanating from other parts of the object are brought into focus in the vicinity of T' such that a complete image of the object is produced between $S'$and T' (obviously, point $S'$is the image of point S). This image could be viewed by projecting it onto a screen placed between points S' and T'. Such an image is termed a real image. Note that the image $S'T'$ would also be directly visible to an observer looking straight at the mirror from a distance greater than the image distance q (since the observer's eyes could not tell that the light-rays diverging from the image were in anyway different from those which would emanate from a real object). According to the figure, the image is inverted with respect to the object, and is also magnified.

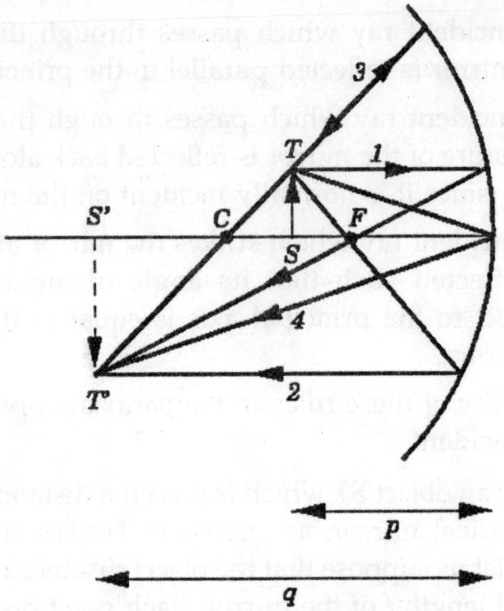

Fig. Formation of a real image by a concave mirror.

Figure shows what happens when the object distance p is less than the focal length f. In this case, the image appears to an observer looking straight at the mirror to be located behind the mirror. For instance, rays emanating from the tip T of the object appear, after reflection from the mirror, to come from a point T' which is behind the mirror. Note that only two rays are used to locate T', for the sake of clarity. In fact, two is the minimum number of rays needed to locate a point image. Of course, the image behind the mirror cannot be viewed by projecting it onto a screen, because there are no real light-rays behind the mirror. This type of image is termed a virtual image. The characteristic difference between a real image and a virtual image is that, immediately after reflection from the mirror, light-rays emitted by the object converge on a real image, but diverge from a virtual image. According to Fig. the image is upright with respect to the object, and is also magnified.

The graphical method described above is fine for developing an intuitive understanding of image formation by concave mirrors, or for checking a calculation, but is a bit too cumbersome for everyday use. The analytic method described below is far more flexible.

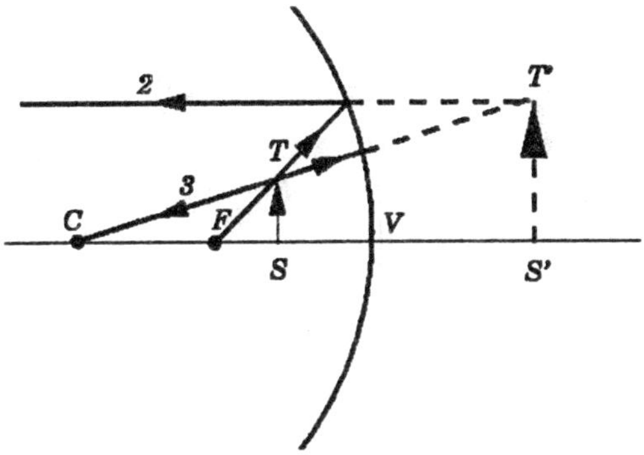

Fig. Formation of a virtual image by a concave mirror.

Consider an object ST placed a distance p in front of a concave mirror of radius of curvature R. In order to find the image $S'T'$ produced by the mirror, we draw two rays from T to the mirror—see Fig. The first, labelled 1, travels from T to the vertex V and is reflected such that its angle of incidence θ equals its angle of reflection. The second ray, labelled 2, passes through the centre of curvature C of the mirror, strikes the mirror at point B, and is reflected back along its own path. The two rays meet at point T'. Thus, $S'T'$ is the image of ST, since point S' must lie on the principal axis.

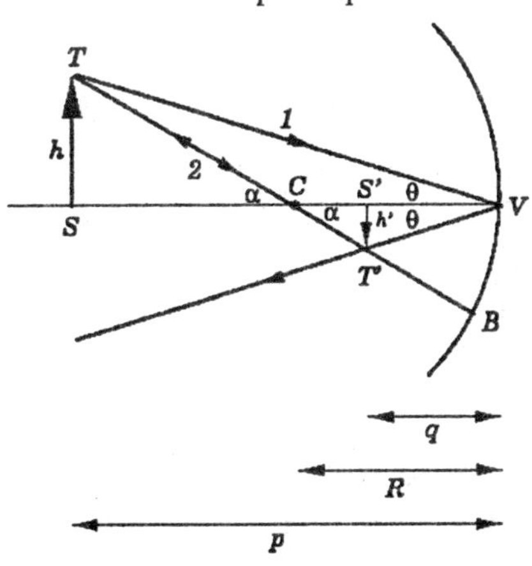

Fig. Image formation by a concave mirror.

In the triangle *STV*, we have $\tan\theta = \frac{h}{p}$, and in the triangle *S'T'V'* we have $\tan\theta = \frac{h'}{q}$, where *p* is the object distance, and *q* is the image distance. Here, *h* is the height of the object, and *h'* is the height of the image. By convention, *h'* is a negative number, since the image is inverted (if the image were upright then *h'* would be a positive number). It follows that

$$\tan\theta = \frac{h}{p} = \frac{h'}{q}.$$

Thus, the magnification *M* of the image with respect to the object is given by

$$M = \frac{h'}{h} = -\frac{q}{p}.$$

By convention, *M* is negative if the image is inverted with respect to the object, and positive if the image is upright. It is clear that the magnification of the image is just determined by the ratio of the image and object distances from the vertex.

From triangles *STC* and *S'T'C*, we have $\tan\alpha = h/(p-R)$ and $\tan\alpha = h'/(R-q)$, respectively. These expressions yield

$$\tan\alpha = \frac{h}{p-R} = -\frac{h'}{R-q}.$$

Equations can be combined to give

$$\frac{-h'}{h} = \frac{R-q}{p-R} = \frac{q}{p}:$$

which easily reduces to

$$\frac{1}{p} + \frac{1}{q} = \frac{2}{R}.$$

This expression relates the object distance, the image distance, and the radius of curvature of the mirror.

For an object which is very far away from the mirror (i.e., $p \to \infty$), so that light-rays from the object are parallel to the principal axis, we expect the image to form at the focal point F of the mirror. Thus, in this case, $q = f$, where f is the focal length of the mirror, and Equation. reduces to

$$0+\frac{1}{f}=\frac{2}{R}.$$

The above expression yields

$$f=\frac{R}{2}.$$

In other words, in the paraxial approximation, the focal length of a concave spherical mirror is half of its radius of curvature. Equations can be combined to give

$$\frac{1}{p}+\frac{1}{q}=\frac{1}{f}.$$

The above expression was derived for the case of a real image. However, as is easily demonstrated, it also applies to virtual images provided that the following sign convention is adopted. For real images, which always form in front of the mirror, the image distance q is positive. For virtual images, which always form behind the mirror, the image distance q is negative. It immediately follows, from Equations. that real images are always inverted, and virtual images are always upright. Table shows how the location and character of the image formed in a concave spherical mirror depend on the location of the object, according to Equations. It is clear that the modus operandi of a shaving mirror, or a makeup mirror, is to place the object (i.e., a face) between the mirror and the focus of the mirror. The image is upright, (apparently) located behind the mirror, and magnified.

Table: Rules for image formation by concave mirrors.

Position of object	Position of image	Character of image
At ∞	At F	Real, zero size
Between ∞ and C	Between F and C	Real, inverted, diminished

At C	At C	Real, inverted, same size
Between C and *F*	Between C and ∞	Real, inverted, magnified
At *F*	At ∞	
Between *F* and *V*	From - ∞ to V	Virtual, upright, magnified
At *V*	At *V*	Virtual, upright, same size

IMAGE FORMATION BY CONVEX MIRRORS

The definitions of the principal axis, centre of curvature *C*, radius of curvature *R*, and the vertex *V*, of a convex mirror are analogous to the corresponding definitions for a concave mirror. When parallel light-rays strike a convex mirror they are reflected such that they appear to emanate from a single point *F* located behind the mirror, as shown in Fig.This point is called the *virtual focus* of the mirror. The focal length *f* of the mirror is simply the distance between *V* and *F*. As is easily demonstrated, in the paraxial approximation, the focal length of a convex mirror is half of its radius of curvature.

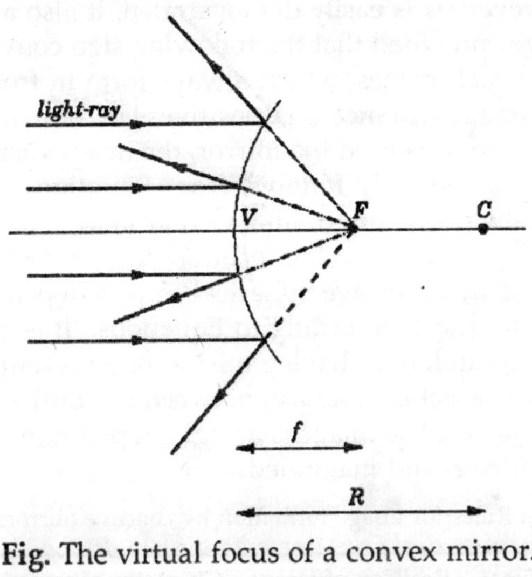

Fig. The virtual focus of a convex mirror.

There are, again, two alternative methods of locating the image formed by a convex mirror. The first is graphical, and the second analytical.

According to the graphical method, the image produced by a convex mirror can always be located by drawing a ray diagram according to four simple rules:

1. An incident ray which is parallel to the principal axis is reflected as if it came from the virtual focus *F* of the mirror.
2. An incident ray which is directed towards the virtual focus *F* of the mirror is reflected parallel to the principal axis.
3. An incident ray which is directed towards the centre of curvature *C* of the mirror is reflected back along its own path (since it is normally incident on the mirror).
4. An incident ray which strikes the mirror at its vertex *V* is reflected such that its angle of incidence with respect to the principal axis is equal to its angle of reflection.

The validity of these rules in the paraxial approximation is, again, fairly self-evident.

In the example shown in Fig. two rays are used to locate the image *S'T'* of an object *ST* placed in front of the mirror. It can be seen that the image is virtual, upright, and diminished.

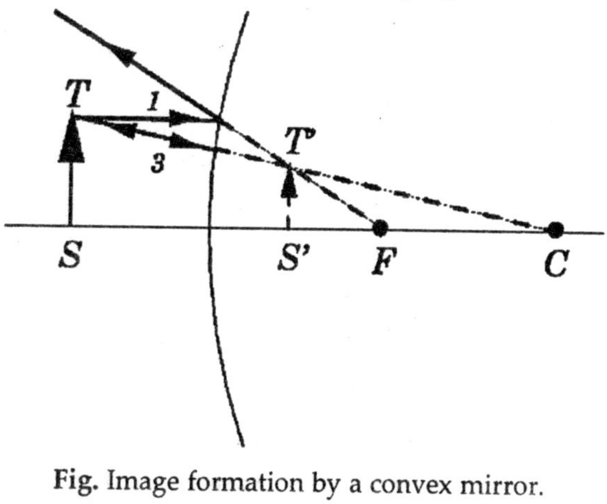

Fig. Image formation by a convex mirror.

As is easily demonstrated, application of the analytical method to image formation by convex mirrors again yields Eq. for the magnification of the image, and Eq. for the location of the image, provided that we adopt the following sign convention. The focal length of a convex mirror is redefined to be *minus* the distance between points and. In other words, the focal length of a concave mirror, with a real focus, is always positive, and the focal length of a convex mirror, with a virtual focus, is always negative. Table shows how the location and character of the image formed in a convex spherical mirror depend on the location of the object, according to Equations. (with).

Table: Rules for image formation by convex mirrors.

Position of object	*Position of image*	*Character of image*
At ∞	At F	Virtual, zero size
Between ∞ and V	Between F and V	Virtual, upright, diminished
At V	At V	Virtual, upright, same size

In summary, the formation of an image by a spherical mirror involves the *crossing* of light-rays emitted by the object and reflected off the mirror. If the light-rays actually cross in front of the mirror then the image is real. If the light-rays do not actually cross, but appear to cross when projected backwards behind the mirror, then the image is virtual. A real image can be projected onto a screen, a virtual image cannot. However, both types of images can be seen by an observer, and photographed by a camera. The magnification of the image is specified by Eq. and the location of the image is determined by Eq. These two formulae can be used to characterize both real and virtual images formed by either concave or convex mirrors, provided that the following sign conventions are observed:

1. The height h' of the image is positive if the image is upright, with respect to the object, and negative if the image is inverted.
2. The magnification M of the image is positive if the image is upright, with respect to the object, and

negative if the image is inverted.

3. The image distance q is positive if the image is real, and, therefore, located in front of the mirror, and negative if the image is virtual, and, therefore, located behind the mirror.
4. The focal length f of the mirror is positive if the mirror is concave, so that the focal point F is located in front of the mirror, and negative if the mirror is convex, so that the focal point F is located behind the mirror.

Note that the front side of the mirror is defined to be the side from which the light is incident.

IMAGE FORMATION BY PLANE MIRRORS

Both concave and convex spherical mirrors asymptote to plane mirrors in the limit in which their radii of curvature R tend to infinity. In other words, a plane mirror can be treated as either a concave or a convex mirror for which $R \to \infty$. Now, if $R \to \infty$, then $f = \pm R/2 \to \infty$, so $1/f \to 0$, and Eq. yields

$$\frac{1}{p} + \frac{1}{q} = \frac{1}{f} = 0,$$

or

$$q = -p.$$

Thus, for a plane mirror the image is *virtual*, and is located as far behind the mirror as the object is in front of the mirror. According to Eq. the magnification of the image is given by

$$M = -\frac{q}{p} = 1$$

Clearly, the image is *upright*, since $M > 0$, and is the *same size* as the object, since $|M| = 1$. However, an image seen in a plane mirror does differ from the original object in one important respect: *i.e.*, left and right are *swapped over*. In other words, a right-hand looks like a left-hand in a plane mirror, and *vice versa*.

THIN LENSES

A lens is a transparent medium (usually glass) bounded by two curved surfaces (generally either spherical, cylindrical, or plane surfaces). As illustrated in Fig. the line which passes normally through both bounding surfaces of a lens is called the *optic axis*. The point *O* on the optic axis which lies midway between the two bounding surfaces is called the *optic centre*.

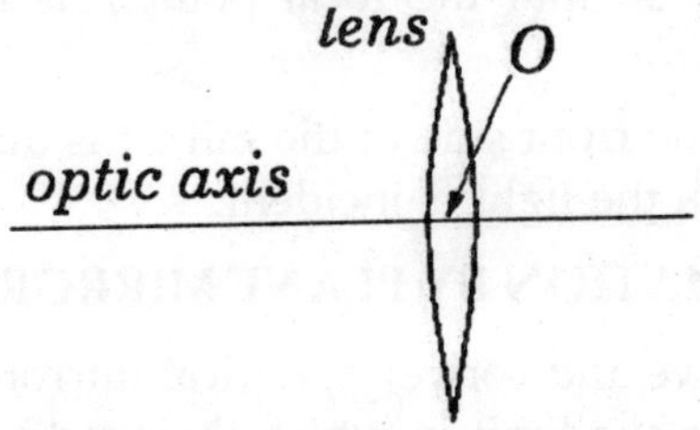

Fig. The optic axis of a lens.

There are two basic kinds of lenses: *converging*, and *diverging*. A converging lens brings all incident light-rays parallel to its optic axis together at a point *F*, behind the lens, called the *focal point*, or *focus*, of the lens. A diverging lens spreads out all incident light-rays parallel to its optic axis so that they appear to diverge from a *virtual focal point F* in front of the lens. Here, the front side of the lens is conventionally defined to be the side from which the light is incident. The differing effects of a converging and a diverging lens on incident light-rays parallel to the optic axis (*i.e.*, emanating from a distant object) are illustrated in Fig.

Lenses, like mirrors, suffer from *spherical aberration*, which causes light-rays parallel to the optic axis, but a relatively long way from the axis, to be brought to a focus, or a virtual focus, *closer* to the lens than light-rays which are relatively close to the axis. It turns out that spherical aberration in lenses can be completely cured by using lenses whose bounding surfaces are *non-spherical*. However, such lenses are more difficult, and, therefore, more expensive, to manufacture than conventional lenses whose bounding surfaces are spherical.

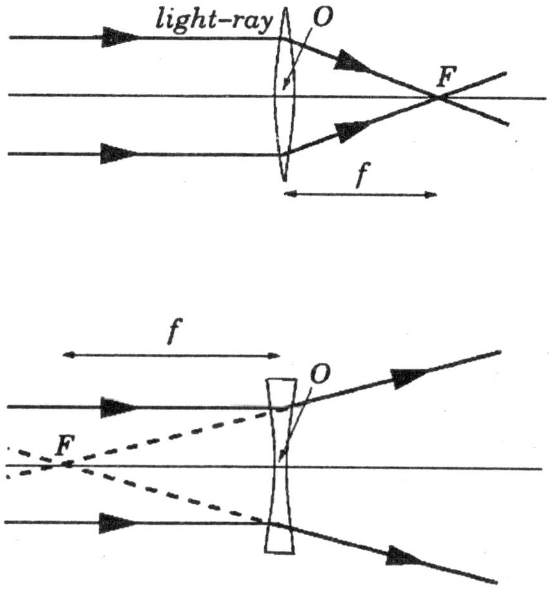

Fig. The focii of converging (top) and diverging (bottom) lens.

Thus, the former sort of lens is only employed in situations where the spherical aberration of a conventional lens would be a serious problem. The usual method of curing spherical aberration is to use *combinations* of conventional lenses (*i.e.*, compound lenses). In the following, we shall make use of the *paraxial approximation,* in which spherical aberration is completely ignored, and all light-rays parallel to the optic axis are assumed to be brought to a focus, or a virtual focus, at the same point *F*. This approximation is valid as long as the radius of the lens is small compared to the object distance and the image distance.

The *focal length* of a lens, which is usually denoted *f*, is defined as the distance between the optic centre *O* and the focal point *F*, as shown in Fig. However, by convention, *converging* lenses have *positive* focal lengths, and *diverging* lenses have *negative* focal lengths. In other words, if the focal point lies behind the lens then the focal length is positive, and if the focal point lies in front of the lens then the focal length is negative.

Consider a conventional lens whose bounding surfaces are *spherical*. Let C_f be the centre of curvature of the front surface, and C_b the centre of curvature of the back surface. The radius of curvature R_f of the front surface is the distance between the optic centre and the point. Likewise, the radius of curvature of the back surface is the distance between points and. However, by convention, the radius of curvature of a bounding surface is *positive* if its centre of curvature lies *behind* the lens, and *negative* if its centre of curvature lies *in front* of the lens. Thus, in Fig. is positive and is negative.

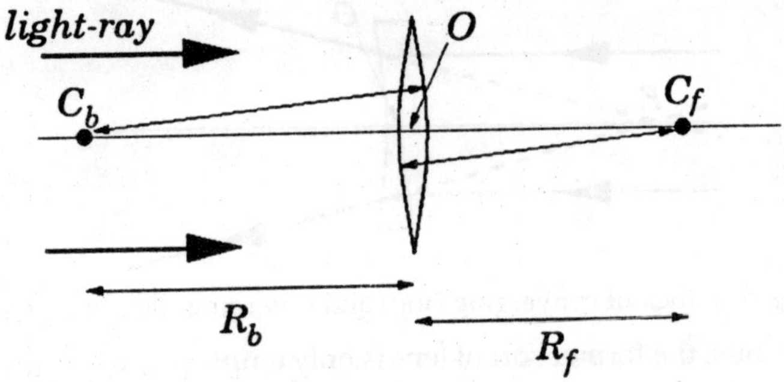

Fig. A thin lens.

In the paraxial approximation, it is possible to find a simple formula relating the focal length f of a lens to the radii of curvature, R_f and R_b, of its front and back bounding surfaces. This formula is written

$$\frac{1}{f}=(n-1)\left(\frac{1}{R_f}-\frac{1}{R_b}\right),$$

where is the refractive index of the lens. The above formula is usually called the *lens-maker's formula,* and was discovered by Descartes. Note that the lens-maker's formula is only valid for a *thin lens* whose thickness is small compared to its focal length. What Eq. is basically telling us is that light-rays which pass from air to glass through a *convex* surface are *focused,* whereas light-rays which pass from air to glass through a *concave* surface are *defocused*. Furthermore, since light-rays are *reversible,* it follows that rays which pass from glass to air through a *convex* surface are *defocused,* whereas rays which pass from air to glass through a *concave* surface are *focused.* Note that the net focusing or defocusing action of a lens is due to the *difference* in the radii of curvature of its two bounding surfaces.

Suppose that a certain lens has a focal length *f*. What happens to the focal length if we turn the lens around, so that its front bounding surface becomes its back bounding surface, and *vice versa*? It is easily seen that when the lens is turned around $R_f \to -R_b$ and $R_b \to -R_f$. However, the focal length of the lens is invariant under this transformation, according to Eq. Thus, the focal length of a lens is the same for light incident from either side. In particular, a converging lens remains a converging lens when it is turned around, and likewise for a diverging lens.

The most commonly occurring type of converging lens is a bi-convex, or double-convex, lens, for which $R_f > 0$ and $R_b < 0$. In this type of lens, both bounding surfaces have a focusing effect on light-rays passing through the lens. Another fairly common type of converging lens is a plano-convex lens, for which $R_f > 0$ and $R_b = \infty$. In this type of lens, only the curved bounding surface has a focusing effect on light-rays. The plane surface has no focusing or defocusing effect. A less common type of converging lens is a convex-meniscus lens, for which $R_f > 0$ and $R_b < 0$, with $R_f < R_b$. In this type of lens, the front bounding surface has a focusing effect on light-rays, whereas the back bounding surface has a defocusing effect, but the focusing effect of the front surface wins out.

The most commonly occurring type of diverging lens is a *bi-concave,* or *double-concave,* lens, for which $R_f > 0$ and $R_b < 0$. In this type of lens, both bounding surfaces have a defocusing effect on light-rays passing through the lens. Another fairly common type of converging lens is a *plano-concave* lens, for which $R_f > 0$ and $R_b = \infty$. In this type of lens, only the curved bounding surface has a defocusing effect on light-rays. The plane surface has no focusing or defocusing effect. A less common type of converging lens is a *concave-meniscus* lens, for which $R_f > 0$ and $R_b > 0$, with $R_f < R_b$. In this type of lens, the front bounding surface has a defocusing effect on light-rays, whereas the back bounding surface has a focusing effect, but the defocusing effect of the front surface wins out.

Figure shows the various types of lenses mentioned above. Note that, as a general rule, converging lenses are thicker at the centre than at the edges, whereas diverging lenses are thicker at the edges than at the centre.

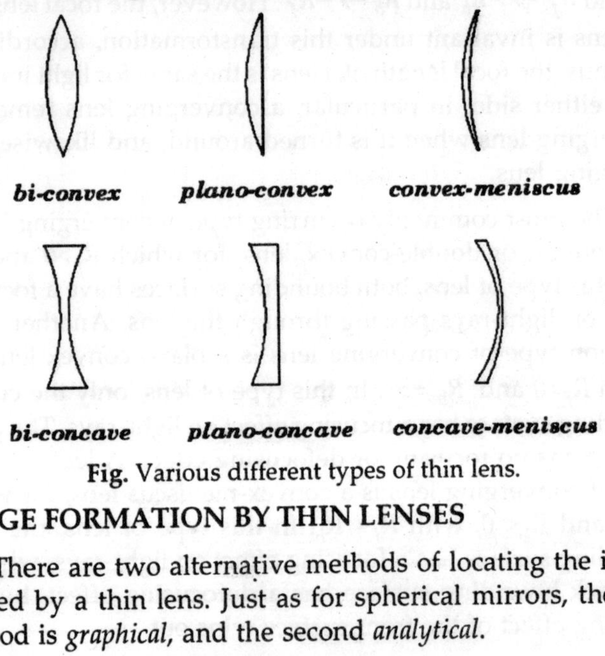

Fig. Various different types of thin lens.

IMAGE FORMATION BY THIN LENSES

There are two alternative methods of locating the image formed by a thin lens. Just as for spherical mirrors, the first method is *graphical*, and the second *analytical*.

The graphical method of locating the image formed by a thin lens involves drawing light-rays emanating from key points on the object, and finding where these rays are brought to a focus by the lens. This task can be accomplished using a small number of simple rules.

Consider a converging lens. It is helpful to define *two* focal points for such a lens. The first, the so-called *image focus*, denoted F_i, is defined as the point behind the lens to which all incident light-rays parallel to the optic axis converge after passing through the lens. This is the same as the focal point F defined previously. The second, the so-called *object focus*, denoted F_o, is defined as the position in front of the lens for which rays emitted from a point source of light placed at that position would be refracted parallel to the optic axis after passing through the lens. It is easily demonstrated that the object focus is as far in front of the optic centre O of the lens as the image focus F_i is behind O. The distance from the optic centre to either focus is, of course, equal to the focal length f of the lens. The image produced by a converging lens can be located using just *three* simple rules:

1. An incident ray which is parallel to the optic axis is refracted through the image focus F_i of the lens.
2. An incident ray which passes through the object focus F_0 of the lens is refracted parallel to the optic axis.
3. An incident ray which passes through the optic centre O of the lens is not refracted at all.

The last rule is only an approximation. It turns out that although a light-ray which passes through the optic centre of the lens does not change direction, it is displaced slightly to one side. However, this displacement is negligible for a thin lens.

Figure illustrates how the image $S'T'$ of an object ST placed in front of a converging lens is located using the above rules. In fact, the three rays, 1-3, emanating from the tip T of the object, are constructed using rules 1-3, respectively. Note that the image is real (since light-rays actually cross), inverted, and diminished.

Consider a diverging lens. It is again helpful to define two focal points for such a lens.

The image focus F_i is defined as the point in front of the lens from which all incident light-rays parallel to the optic axis appear to diverge after passing through the lens. This is the same as the focal point F defined earlier.

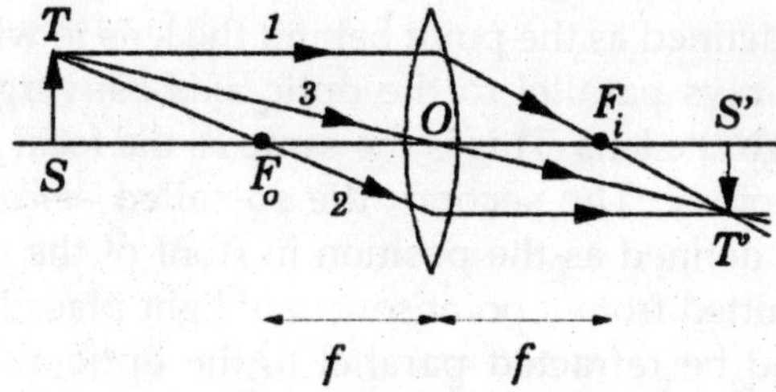

Fig. Image formation by a converging lens.

The object focus is defined as the point behind the lens to which all incident light-rays which are refracted parallel to the optic axis after passing through the lens appear to converge. Both foci are located a distance f from the optic centre, where f is the focal length of the lens. The image produced by a diverging lens can be located using the following three rules:

- An incident ray which is parallel to the optic axis is refracted as if it came from the image focus F_i of the lens.
- An incident ray which is directed towards the object focus of the lens is refracted parallel to the optic axis.
- An incident ray which passes through the optic centre of the lens is not refracted at all.

Figure illustrates how the image $S'T'$ of an object ST placed in front of a diverging lens is located using the above rules. In fact, the three rays, 1-3, emanating from the tip T of the object, are constructed using rules 1-3, respectively. Note that the image is virtual (since light-rays do not actually cross), upright, and diminished.

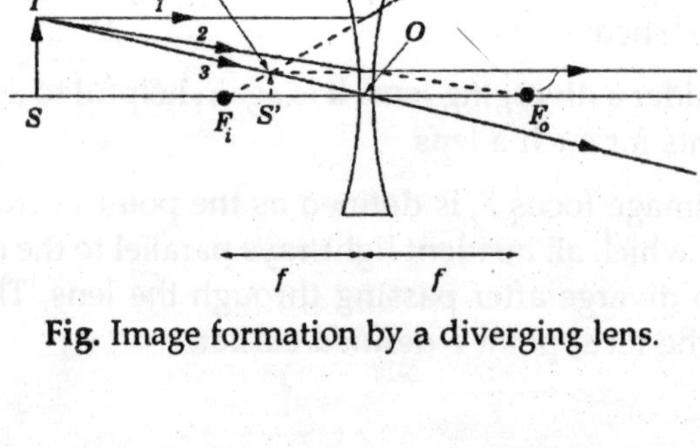

Fig. Image formation by a diverging lens.

Let us now investigate the analytical method. Consider an object of height h placed a distance p in front of a converging lens. Suppose that a real image of height h' is formed a distance q behind the lens. As is illustrated in Fig. the image can be located using rules 1 and 3, discussed above.

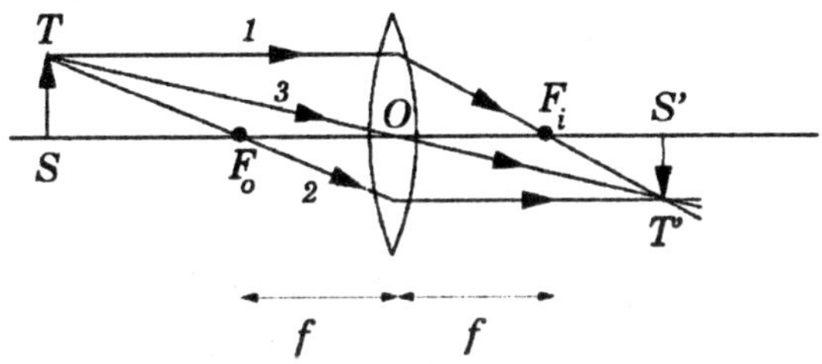

Fig. Image formation by a converging lens.

Now, the right-angled triangles SOT and $S'OT'$ are similar, so

$$\frac{-h'}{h} = \frac{OS'}{OS} = \frac{q}{p}.$$

Here, we have adopted the convention that the image height is *negative* if the image is *inverted*. The magnification of a thin converging lens is given by

$$M = \frac{h'}{h} = -\frac{q}{p}.$$

This is the same as the expression for the magnification of a spherical mirror. Note that we are again adopting the convention that the magnification is *negative* if the image is *inverted*.

The right-angled triangles OPF and $S'T'F$ are also similar, and so

$$\frac{S'T'}{OP} = \frac{FS'}{OF},$$

or

$$\frac{-h'}{h} = \frac{q}{p} = \frac{q-f}{f}.$$

The above expression can be rearranged to give

$$\frac{1}{p}+\frac{1}{q}=\frac{1}{f}.$$

Note that this is exactly the same as the formula relating the image and object distances in a spherical mirror.

Although formulae and were derived for the case of a real image formed by a converging lens, they also apply to virtual images, and to images formed by diverging lenses, provided that the following sign conventions are adopted. First of all, as we have already mentioned, the focal length f of a *converging* lens is *positive*, and the focal length of a *diverging* lens is *negative*. Secondly, the image distance q is *positive* if the image is *real*, and, therefore, located *behind* the lens, and *negative* if the image is *virtual*, and, therefore, located *in front* of the lens. It immediately follows, from Eq. that *real* images are always *inverted*, and *virtual* images are always *upright*.

Table shows how the location and character of the image formed by a converging lens depend on the location of the object. Here, the point V_0 is located on the optic axis two focal lengths in front of the optic centre, and the point V_i is located on the optic axis two focal lengths behind the optic centre. Note the almost exact analogy between the image forming properties of a converging lens and those of a concave spherical mirror.

Table: Rules for image formation by converging lenses.

Position of object	Position of image	Character of image
At $+\infty$	At F	Real, zero size
Between $+\infty$ and V_0	Between F and V_i	Real, inverted, diminished
At V_0	At V_i	Real, inverted, same size
Between V_0 and F	Between V_i and $-\infty$	Real, inverted, magnified
At F	At $-\infty$	
Between F and 0	From $+\infty$ to 0	Virtual, upright, magnified
At 0	At 0	Virtual, upright, same size

Table shows how the location and character of the image formed by a diverging lens depend on the location of the object. Note the almost exact analogy between the image forming properties of a diverging lens and those of a convex spherical mirror.

Table: Rules for image formation by diverging lenses.

Position of object	Position of image	Character of image
At ∞	At F_i	Virtual, zero size
Between ∞ and 0	Between F_i and 0	Virtual, upright, diminished
At 0	At 0	Virtual, upright, same size

Finally, let us reiterate the sign conventions used to determine the positions and characters of the images formed by thin lenses:

1. The height h' of the image is positive if the image is upright, with respect to the object, and negative if the image is inverted.
2. The magnification M of the image is positive if the image is upright, with respect to the object, and negative if the image is inverted.
3. The image distance q is positive if the image is real, and, therefore, located behind the lens, and negative if the image is virtual, and, therefore, located in front of the lens.
4. The focal length f of the lens is positive if the lens is converging, so that the image focus F_i is located behind the lens, and negative if the lens is diverging, so that the image focus F_i is located in front of the lens.

Note that the front side of the lens is defined to be the side from which the light is incident.

Chapter 6

The Optical Field

Optical fields lie on the electromagnetic spectrum in the frequency range from 10^{14} to 10^{16} Hertz. The hard ultraviolet and x-ray regimes lie at frequencies above the optical region and the mid and far infrared and microwave regimes lies at frequencies below. The most distinguishing aspect of optical fields is that they are visible to the human eye. As there is little fundamental difference between physical phenomena observed for the visual spectrum and phenomena observed in the near infrared and ultraviolet, however, these regions are often included in the optical regime.

Two characteristics distinguish optics from other electromagnetic technologies. First, optical wavelengths (0.1 to 10 microns) are small compared to typical aperture sizes and measurement distances, meaning that optical fields are usually considered in far-field and weakly diffracting limits. Second, optical fields are generated and detected by atomic-scale quantum interactions rather than by antennas or resonators. The first characteristic leads one to analyse optical propagation under approximations to the Maxwell equations appropriate to the short wavelength regime. In many cases, one can neglect diffraction and consider optical excitations to consist of straight "beams." In such cases, one uses geometrical analysis based on ray tracing to understand the general behaviour of optical systems. The second characteristic means that optical field generation and detection is not amenable to direct artificial control and that optical fields are susceptible to different and more profound sources of noise and uncertainty than other regions of the spectrum.

Optical fields are central to three technologies: sensing, communication and illumination. Sensing is the oldest of these, since cavemen had eyes, but fire for illumination came not much later and fire for smoke signal communication soon followed. (The earliest long distance communication might have been "Urg's hut is on fire.") In the modern world, optical sensing spans a range from microscopy to astronomical telescopy and includes still, cinematic and video photography, lidar, reflectometry, and a number of other applications. Optical sensing is increasingly dominated by electronic sensors and digital processing. Despite the long history of smoke signals, lighthouses, lanterns and signal lights, optical communications has bloomed as an industry only in the past 20 years. The key to this development has been the explosion in demand for data transmission, the development of astonishingly low loss optical fiber and the development of semiconductor laser sources and detectors. Illumination has a longer history as a major industry than either sensing or communications. Illumination today includes the range from flashlights to laser tools for surgery. In view of the greater potential for research and engineering at the higher tech end of the scale, modern engineering courses tend to emphasize lasers over lightbulbs. There are other optical technologies which are difficult to classify in these three areas, for example, optical data storage mixes sensing, communications and illumination. This course covers sensing applications of optical fields. In the UIUC ECE department ECE 371 GP covers communications applications and ECE 355 covers lasers.

Just as there are three major optical applications, there are three main questions to be resolved in analyzing optical systems: how is the field generated?, how does it propagate? and how is it detected? These questions play a role in all three technologies, the significance and focus varies. Communications system development focuses primarily on modulation and detection technologies. Illumination system development focuses primarily on source issues. Sensor systems are most concerned with how the field propagates from the object or system being sensed to the detection system. While the nature of the illuminating field (e.g. its polarization, spectral and coherence properties) and the fidelity of the detection system are important, the enabling issue for sensing is how to detect information which can be back propagated to identify the object.

The primary reason for the emphasis on field propagation in sensing systems is that the most sensors detect parallel spatial information. In contrast, the spatial pattern generated by most illumination systems contain little information and communication systems limit the field to a single spatial mode or a very small number of modes in a guiding fiber. Since the spatial pattern of the field is information rich in sensing applications, it is important to understand how this information is transformed by the spatial distortions of propagation. The propagation of the optical field is described by Maxwell's equations. Since, as will be discussed later in the course, the optical field itself is not directly detectable, statistical correlations and intensity measures must supplement Maxwell's equations to fully understand propagation.

MAXWELL'S EQUATIONS

Maxwell's equations in differential form are

$$\nabla \times \mathbf{E} = -\frac{\partial}{\partial t}\mathbf{B}$$

$$\nabla \times \mathbf{H} = \mathbf{J} + \frac{\partial}{\partial t}\mathbf{D}$$

$$\nabla \cdot \mathbf{D} = \rho$$

$$\nabla \cdot \mathbf{B} = 0$$

is the electric field, is the electric displacement, B is the magnetic induction, H is the magnetic field, J is the current density and is the charge density. The first equation is

Faraday's Law and expresses the experimental fact that a time varying magnetic field induces an electromotive force. The second equation is Ampere's Circuital Law and expresses the experimental fact that a current or time varying electric field induces a magnetic field. The last two equations are Gauss's Law and express the experimental facts of Coloumb's law and charge conservation.

The field variables are related by the material equations

$$\mathbf{D} = \varepsilon_o \mathbf{E} + \mathbf{P}$$

$$\mathbf{B} = \mu_o \mathbf{H} + \mathbf{M}$$

where P is the *polarization* of material and M is the *magnetization*. In most optical materials, $M = 0$ and P is a function of E. The simplest and most common case is

$$\mathbf{P} = \varepsilon_o \chi_e \mathbf{E}$$

Then

$$\mathbf{D} = \varepsilon \mathbf{E}$$

where

$$\varepsilon = \varepsilon_o \varepsilon_r$$

and

$$\varepsilon_r = 1 + \chi_e$$

As a result of the high frequencies and small wavelengths of optical fields, charge interactions with the optical field involve quantum mechanical effects which cannot accurately be analysed by continuous models. In optical diffraction problems, is always neglected. In principle, J is also neglected, although the current density can be used formally to add a loss component to the pemitivitty,. We also note that the vectors E and D need not be colinear, meaning that may in general be tensor valued. A medium in which is (not) tensor valued is called (isotropic) anisotropic.

THE WAVE EQUATION

Using the material relations, we substitute in the Maxwell equations to find the wave equations in isotropic media.

$$\nabla \times \nabla \times \mathbf{E} = -\mu_o \varepsilon \frac{\partial^2}{\partial t^2} \mathbf{E}$$

$$\nabla \times \nabla \times \mathbf{H} = -\mu_o \varepsilon \frac{\partial^2}{\partial t^2} \mathbf{H}$$

The equations are reduced to a simpler form by the vector identity

$$\nabla \times \nabla \times \mathbf{A} = \nabla(\nabla \cdot \mathbf{A}) - \nabla^2 \mathbf{A}$$

From the Gauss's Law we know that

$$\nabla \cdot (\varepsilon \mathbf{E}) = \mathbf{E} \cdot \nabla \varepsilon + \varepsilon \nabla \cdot \mathbf{E} = 0,$$

where we have assumed for the moment ε that is scalar valued. Thus,

$$\nabla \cdot \mathbf{E} = -\mathbf{E} \cdot \nabla \log(\varepsilon)$$

and

$$\nabla^2 \mathbf{E} + \nabla(\mathbf{E} \cdot \nabla \log(\varepsilon)) - \mu_o \varepsilon \frac{\partial^2}{\partial t^2} \mathbf{E} = 0$$

A medium in which $\nabla \varepsilon = 0$ (the permittivity is spatially constant) is *homogeneous*. Most common materials are approximately homogeneous. Media in which this is not the case include optical fiber, graded index lenses, and volume holograms.

In homogeneous isotropic media, the wave equations are

$$\nabla^2 \mathbf{E} - \mu_o \varepsilon \frac{\partial^2}{\partial t^2} \mathbf{E} = 0$$

$$\nabla^2 \mathbf{H} - \mu_o \varepsilon \frac{\partial^2}{\partial t^2} \mathbf{H} = 0$$

SOLUTIONS TO THE WAVE EQUATION

Fields satisfying the wave equations are 4D disturbances in space-time. Since a wave disturbance propagates without dissipation, there are no spatially bounded solutions to the wave equation in source and absorber free space. Pulse packets in localized in space-time are potential solutions, but such solutions diffractively dissipate on propagation. There are no simple exact solutions of finite energy.

Plane waves are the simplest solutions to the wave equation. Plane waves are unbounded and constant in two dimensions and propagate along a third dimension. Assuming that the constant dimensions are x and y, the derivative of the fields vanish along these dimension and the wave equation for the electric field reduces to

$$\frac{\partial^2}{\partial z^2}E_x - \mu_o\varepsilon\frac{\partial^2}{\partial t^2}E_x = 0$$

where we have chosen to write the equation only for one polarization component of the electric field. Note that for the plane wave to satisfy Gauss's law the z component of the field must be zero.

Eqn. has many possible solutions, the simplest being

$$E_x(z) = f\left(t \pm \frac{z}{v_p}\right)$$

where $v_p = \frac{1}{\sqrt{\mu_0 E}}$ and $f(t)$ is any twice differentiable function. This solution is a wave propagating in the negative (plus) z direction if the plus (negative) sign is selected.

We consider some simple animations to provide some feeling for solutions in the optical domain. The first example is an pulse with a centre wavelength of 0.5 microns, in the green/blue region of the optical spectrum. The pulse oscillates at a centre frequency of at approximately 0.610^{15} Hz. This particular pulse has a Gaussian envelop. The pulse width of the Gaussian is 10 femtoseconds. The field amplitude of the pulse can be written

$$f(t) = \cos\left(0.610^{15}t\right)\exp\left(-\left(\frac{t}{10^{-14}}\right)^2\right).$$

The animation below shows the field amplitude as a function space along the axis of propagation. Each frame in the animation represents a time shift of 20 femtoseconds. The pulse propagates to the right at a velocity of 0.3 microns/ femtosecond.

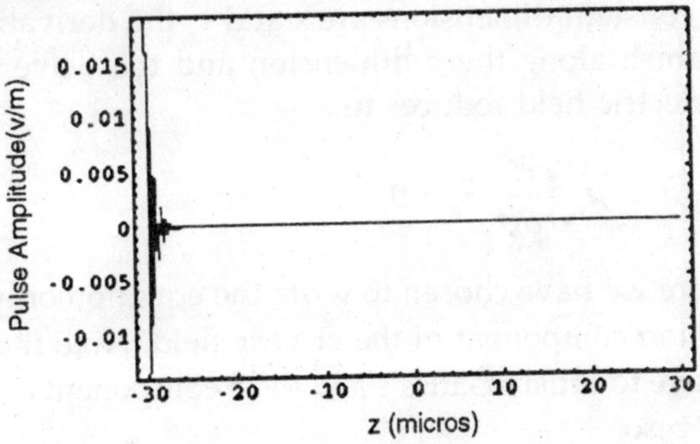

The solution $f(t)$ need not be spatially confined. For Fourier analysis of fields, unbounded harmonic solutions such as $f(x) = \cos\left(0.610^{15} t\right)$ are popular. For such fields, one perceives wave propagation in the phase of the excitation. An animation of this field propagating in the plus z direction is shown below:

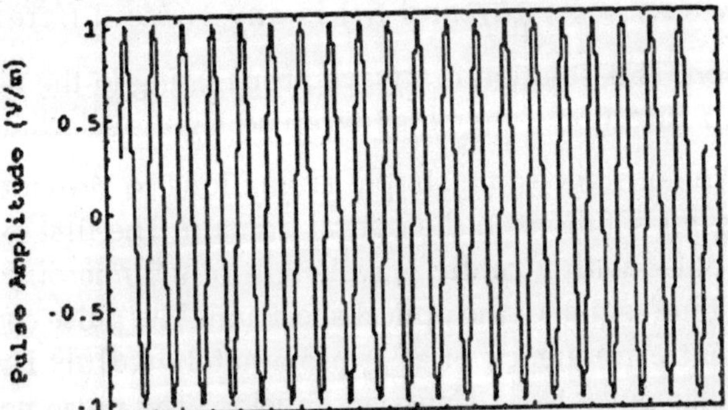

OPTICAL SENSORS

As mentioned in the introductory note on coherence theory, the nature of optical detectors plays a key role in the statistical analysis of optical systems. Detection of spatially distributed fields relied exclusively on photochemical materials or the photoelectric effect until the invention of CCD arrays in the mid'70s. A practical discussion of how these devices function is useful in considering noise sources.

ANALYSIS OF FRESNEL DIFFRACTION

This note shows a couple of examples of the calculation of Fresnel transformations. Calculation of Fresnel diffraction presents a number of interesting challenges, some are pointed out by the example here, some you will discover on your own. Given that you are likely to encounter erroneous results, the important thing is to keep your intuition working.

Our first example is based on the input field

$$f(x,y) = \exp\left(-pi\frac{(x^2+y^2)}{A^2}\right)$$

Consider

$$f(x,y) = \left(\exp\left(-100pi\frac{(x+2A)^2}{A^2}\right) + \exp\left(-100pi\frac{(x+A)^2}{A^2}\right) + \exp\left(-100pi\frac{(x)^2}{A^2}\right)\right.$$
$$\left. + \exp\left(-100pi\frac{(x-A)^2}{A^2}\right) + \exp\left(-100pi\frac{(x-2A)^2}{A^2}\right)\right) \times$$
$$\left(\exp\left(-100pi\frac{(y+2A)^2}{A^2}\right) + \exp\left(-100pi\frac{(y+A)^2}{A^2}\right) + \exp\left(-100pi\frac{(y)^2}{A^2}\right)\right.$$
$$\left. + \exp\left(-100pi\frac{(y-A)^2}{A^2}\right) + \exp\left(-100pi\frac{(y-2A)^2}{A^2}\right)\right)$$

NUMERICAL ANALYSIS OF COHERENT IMAGING

Coherent Imaging

The transfer function for a coherent imaging system is

$$\tilde{H}(\mu v) = P(-\lambda d_i \mu, -\lambda d_i v),$$

where $P(x,y)$ describes the transmittance of the output pupil of the imaging system. The impulse response of the imaging system is the impulse response of the transfer function:

$$\tilde{h}(x_i y_i) = \iint (\lambda d_i \mu, \lambda d_i \nu) e^{j2\pi(x\mu + y\nu)} d\mu d\nu$$

We find the image plane field, $g(x_i, y_i)$, from the object plane field, $f(x_o, y_o)$, using

$$g(x_i, y_i) = \iint \tilde{h}(x_i - \tilde{x}_o, y_i - \tilde{y}_o) \frac{1}{M^2} f\left(\frac{\tilde{x}_o}{M}, \frac{\tilde{y}_o}{M}\right) d\tilde{x}_o d\tilde{y}_o$$

where

$$M = \frac{d_i}{d_o}.$$

INCOHERENT IMAGING

Incoherent imaging is a linear system in the input intensity. The impulse response for an incoherent imaging system is equal to the squared modulus of the field impulse response for the corresponding coherent system. The transfer function for the incoherent imaging system is thus the autocorrelation of the coherent transfer function. As discussed in the note on coherent imaging, the transfer function for a coherent system described by the exit pupil function $P(x,y)$ is

$$\tilde{H}(u, v) = P(-\lambda d_i u, -\lambda d_i v)$$

The first section of this document analyses the incoherent transfer function and impulse response for various apertures.

Incoherent Transfer Functions and Impulse Responses

The normalized transfer function for an incoherent system is called the OTF, or optical transfer function. The Fourier transform of the OTF is the point spread function (PSF) of the imaging system. The OTF satisfies the following constraints:

1. $\mathcal{H}(0,0) = 1$
2. $\mathcal{H}(-f_x, -f_y) = \mathcal{H}^*(f_x, f_y)$
3. $|\mathcal{H}(f_x, f_y)| \leq |\mathcal{H}(0,0)|$

Depth of Field

The impulse response for coherent image away from the focal plane is

$$h(x_i,y_i,x_o,y_o)=\frac{1}{\lambda^2 d_o d_i}\exp\left[\frac{j\pi}{\lambda d_i}(x_i^2+y_i^2)\right]\exp\left[\frac{j\pi}{\lambda d_o}(x_o^2+y_o^2)\right]\times$$

$$\iint P(x,y)\exp\left[\frac{j\pi}{\lambda}\left(\frac{1}{d_i}+\frac{1}{d_o}-\frac{1}{f}\right)(x^2+y^2)\right]\times$$

$$\exp\left[-\frac{j2\pi}{\lambda}\left(\frac{x_i}{d_i}+\frac{x_o}{d_o}\right)x\right]\exp\left[-\frac{j2\pi}{\lambda}\left(\frac{y_i}{d_i}+\frac{y_o}{d_o}\right)y\right]dxdy$$

For the incoherent imaging system, the impulse response is simply $|h(x_i,y_i,x_o,y_o)|^2$. Squaring eliminates the phase factors in front of the integral and makes the impulse response shift invariant under the coordinate transformations discussed in the note on coherent imaging. The effective coherent transfer function for this transformation is

$$\tilde{H}(u,v)=P(-\lambda d_i u,-\lambda d_i v)\exp\left[j\pi\lambda d_i^2\left(\frac{1}{d_i}+\frac{1}{d_o}-\frac{1}{f}\right)(u^2+v^2)\right]$$

Using this transfer function in the code above, we can put in a variety of values for d_i (assuming that d_o is fixed at 1 cm and f is fixed at 5 mm). The animation below shows the point spread function for a 1 mm aperture as the image position moves 100 microns toward the lens from the focal plane.

DIFFRACTION FREE BEAMS

Diffraction free beam is invariant with respect to diffraction from one plane to the next. A plane wave is an example of such a field. It turns out that plane waves are not unique in this respect and that unbounded solutions to the Maxwell equations exist which localize their energy along the optical axis more effectively than plane waves. Of course, it is not possible for a beam passing through a finite aperture to avoid diffraction. The Bessel beam fields described here propagate without diffraction through unbounded apertures, but the energy of these beams is not finite and the beam does diffract after passing through an aperture.

One can easily verify that the field

$$E(\bar{r}) = e^{-j\beta z}\int_0^{2\pi} A(\phi)\exp\left[j\alpha(x\cos\phi + y\sin\phi)\right]d\phi$$

where $A(\phi)$ is an arbitrary complex function of ϕ and $\beta^2 + \alpha^2 = \left(\frac{\omega}{c}\right)^2$, is a solution to the Helmholtz equation.

The particular solution $A(\phi)=\frac{1}{2\pi}$ can be integrated to yield the non-diffracting beam

$$E(\bar{r}) = e^{-j\beta z}\int_0^{2\pi} \exp\left[j\alpha(x\cos\phi + y\sin\phi)\right]\frac{d\phi}{2\pi}$$

$$= e^{-j\beta z}J_0(\alpha\rho)$$

To see why this is a nondiffracting beam, consider its representation in k-space. The field consists of plane wave components drawn from a ring around the optical axis, as sketched below. Since each plane wave component experiences the same phase delay along the axis as the field propagates, all the components stay in phase and the transverse structure of the field is unchanged with propagation. It is not necessary to use the entire ring to achieve this objective, you might want to consider fields drawn from partial segments on your own. Two points on opposite sides of the ring, for example, would form a nondiffracting cosine transverse pattern.

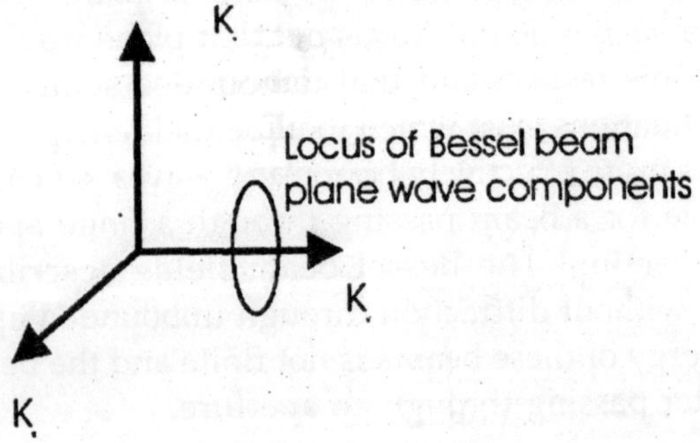

IMAGING AND COHERENCE

Aside from the van Cittert-Zernike theorem, MW does not consider image formation. A primary goal of this course is to develop a detailed understanding of spatial pattern formation and imaging with optical fields. GS has a much better analysis of imaging. We will spend the last 2-3 weeks of the term discussing image formation and the information capacity of imaged fields using GS, references and notes posted here.

OPTICS AND INFORMATION

Of all physical phenomena, optics allows by far the highest capacity, greatest bandwidth and most effective means of information transfer. An optical fiber can transfer information at bandwidths 3-4 orders of magnitude beyond the fastest copper wire and optical imaging systems surpass the bandwidths and parallelism of ultrasonic and radio wave imaging systems by even greater margins. Given these vast information transfer capabilities, analysis of information capacity and flow often lies at the heart of optical system analysis. This note briefly reviews the quantitative concept of information and briefly describes how it arises in optical analysis. Sampling theorems, which lie at the heart of information analysis in analog systems, are discussed at the conclusion of the note.

QUANTITATIVE MEASUREMENT OF INFORMATION

Information is stored in systems, transferred over channels and detected by measurements. A system might be "my crazy old aunt Beatrice, sitting on a sofa in the living room." Let's call this system S. Information from S might vary from details such as "Aunt Bea's knees are dirty" to "there is a nanogram of plutonium on the third gray hair back from Aunt Bea's left earlobe.". Leaving aside the philosophical question as to whether or not system S stores an infinite amount of information, the quantity of information about S which can be transferred over a channel or detected by a measurement is always finite. Given that this is the case, we often consider the communication channel or the measurement device to be part of system S and define a quantitative measure for the information capacity of S. A suitable definition is:

The information capacity, H, of a system S is the minimum number of bits necessary to completely specify the state of S.

The number of yes/no questions that could be answered about S is an upper bound on H. For example the answer to the question "Are Aunt Bea's knees dirty?" could convey one bit of information. The number of yes/no questions is an upper bound on H because the answers to some questions may be correlated or implicit in the nature of the system. For example, if Aunt Bea's knees are always dirty, the answer to the question may convey no information.

The definition of the information capacity makes comparisons between systems possible and makes information a universal concept. For example, if we can fully specify the state of S using one megabit, than the information capacity of S is equivalent to a one megabit computer memory and the systems themselves are in some sense equivalent. It is important to stress, however, that there is a difference between specifying the state of a system and describing the state of a system. The number of bits needed to fully describe Aunt Bea sitting on the sofa may greatly exceed the number needed to specify the state, particularly if describing Aunt Bea requires the use of poetry. To further see the distinction, consider the system "the book I checked out of the library." If the library contains 32 books, this system contains 5 bits of information, which is the number of bits needed to specify which book I checked out. On the other hand the system "a 200 page book" may contain between one and several hundred megabytes of information, depending on whether or not the book is all text or contains figures.

The library book example leads to a simple means of determining the information capacity of a system: If a system S exists with equal probability in N different states, than the information capacity, H, of S is the base 2 logarithm of N. As an example of the use of this definition to calculate information, consider the system "images in the movie Citizen

Kane." Citizen Kane runs 119 minutes at 24 frames per second, so this system contains log(119*24*60) or 17.4 bits. In contrast, the system "a binary image on black and white film," contains a million bits if we assume that there are a million resolvable spots in on the film.

INFORMATION IN ANALOG SYSTEMS

In digital systems, the quantity information is equal to the number of bits available and is thus quite precise. In analog systems, such as an imaging system, information is more difficult to quantify. Noise and measurement resolution must be considered in analog systems and play key roles in determining information capacity. In analyzing the information capacity of imaging systems, it is often desirable to separate the impact of noise and measurement resolution, which are determined by the illumination source and the detector, from the impact of diffraction and spatial apertures. Analysis of field propagation (diffraction) and boundary conditions (apertures) determines the number of electromagnetic modes which propagate through an imaging system. The temporal bandwidth and noise characteristics of the illumination source and the detector and digitization resolution on detection determine the quantity of information each mode can carry.

In analyzing the field in imaging systems, one focuses on the number of degrees of freedom. Each degree of freedom is a variable on which information can be encoded. For example, the magnitude and phase of a particular electromagnetic mode might constitute a degree of freedom and the number of available modes might constitute the number of degrees of freedom. Discussions of degrees of freedom allow one to estimate the potential information capacity of an analog system without fully considering the impact of noise and measurement. A system with N degrees of freedom in which each degree of freedom could be independently placed and measured in r different states has an information capacity of

$$H = \log\left(r^N\right) = N\log(r)$$

Of course, it is critically important that the noise and measurement processes be such that r is greater than 2. In general, r is rarely large enough to play a major role in increasing the information capacity. Since the information is linear in N and linear in the log of r, it is almost always more useful to increase N rather than r.

As mentioned above, analysis of the degrees of freedom of an optical system may be considered from an electromagnetic perspective in terms of the modes of the system. A *mode* is a field distribution which satisfies both the Maxwell equations and the boundary conditions for a particular geometry. Usually, one chooses to work with a set of modes which are *orthogonal*, which means that their inner product is zero, and which are complete, which means that any field satisfying the Maxwell equations in the geometry of interest can be described by a superposition of modes drawn from the set. In geometries where the field is confined to a limited range of space or spatial frequencies, a set of orthogonal and complete modes is generally discrete. In this case, the spatially continuous field distribution can be represented by a discrete (although potentially infinite) list of mode amplitudes.

THE SAMPLING THEOREM

Similar methods for representing continuous functions using discrete lists arise in linear systems analysis in the context of sampling theorems. Consider as an example a square integrable bandlimited function $f(x,y)$. We focus on 2D functions because our main focus will be on fields measured across 2D planes, such as image planes. By *bandlimited,* we mean that the Fourier transform of f is vanishes outside a certain range. We might express f as

$$f(x,y) = \int_{-B_x}^{B_x} \int_{-B_y}^{B_y} F(u,v) \exp\left[j2\pi(ux+vy)\right] du dv$$

The Fourier transform of f, $F(u,v)$, is limited to a square region of area $4B_xB_y$ in the Fourier plane. A density plot of an example form of $F(u,v)$ is sketched below:

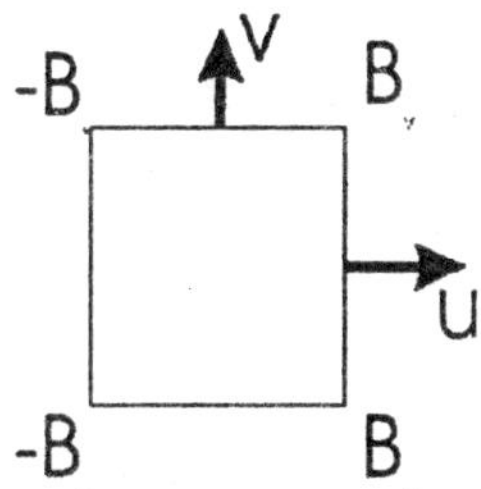

Consider a function $F_p(u,v)$, which is a periodic tiling of the Fourier plane with copies of $F(u,v)$, as sketched below.

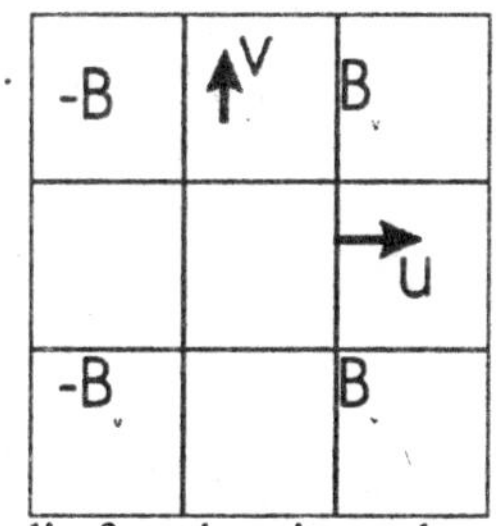

Since $F_p(u,v)$ is a periodic function, it can be expressed as the Fourier series

$$F_p(u,v)=\sum_{n=-\infty}^{\infty}\sum_{m=-\infty}^{\infty} a_{nm}\exp\left[-j\pi\left(\frac{nu}{B_x}+\frac{mv}{B_y}\right)\right].$$

Over the range of $F(u,v)$, $F(u,v) = F_p(u,v)$. This means that

$$f(x,y)=\int_{-B_x}^{B_x}\int_{-B_y}^{B_y} F_p(u,v)\exp\left[j2\pi(ux+vy)\right]du\,dv$$

$$=\sum_{n=-\infty}^{\infty}\sum_{m=-\infty}^{\infty} a_{nm}\int_{-B_x}^{B_x}\int_{-B_y}^{B_y}\exp\left[j2\pi\left(u\left(x-\frac{n}{2B_x}\right)+v\left(y-\frac{m}{2B_y}\right)\right)\right]du\,dv$$

$$=\sum_{n=-\infty}^{\infty}\sum_{m=-\infty}^{\infty} a_{nm}\,\mathrm{sinc}\left(2B_x x-n\right)\mathrm{sinc}\left(2B_y y-m\right)$$

where

$$\mathrm{sinc}(x)=\frac{\sin(\pi x)}{\pi x}$$

Considering the case $x=\frac{n}{2B},y=\frac{m}{2B_y}$, we find $f\left(\frac{n}{2B_x},\frac{m}{2B_y}\right)=a_{nm}$,

which yields the Whittaker-Shannon sampling theorem,

$$f(x,y)=\sum_{n=-\infty}^{\infty}\sum_{m=-\infty}^{\infty} f\left(\frac{n}{2B_x},\frac{m}{2B_y}\right)\text{sinc}\left(2B_x x-n\right)\text{sinc}\left(2B_y y-m\right)$$

As discussed above, the sampling theorem means that the continuous function $f(x,y)$ can be fully represented by the discrete set of samples

$$f\left(\frac{n}{2B_x},\frac{m}{2B_y}\right)$$

In cases of particular interest, we focus on the value of $f(x,y)$only over some finite aperture. If we leave $f(x,y)$ unconstrained outside this aperture, only a finite number of samples is needed to characterize the function. If the extent of the aperture of interest is (-X,X) and (-Y,Y), the number of samples needed is $16B_xB_yXY$. This number of samples corresponds to the number of degrees of freedom in the system "$f(x,y)$ over the aperture of interest." As discussed above, the information capacity of this function is proportional to this number of degrees of freedom.

Geometrical optics: The domain in which optical waves can be treated as simple straight-line rays and the effect of optical elements can be modeled using simple geometrical constructions. There are a number of requirements for these approximations to be an accurate description of an optical system. The most important is that $\pi/D \ll 1$, where π is the wavelength of the light signal and D is the size of the object, lens, aperture, etc. that is being studied. This requirement is almost always satisfied for visible light interacting with everyday-sized objects (which is why Newton did not find a contradiction from everyday experience when he stated that light was probably a stream of particles).

There are a number of other requirements that must be fulfilled in some cases. These requirements will become important in the studies of lenses and curved mirrors.

PROPERTIES OF SHADOWS

Shadow: produced on a screen behind the object when light from a source is blocked (either completely or only partially) by the object. In the case where geometrical optics is appropriate, the shadow can be constructed using simple straight-line rays.

A clear unambiguous shadow requires a "point" source of light – something that either is very small compared to the other dimensions of the experiment or else is very far away. The geometrical construction of the shadow is particularly simple in this case, since all of the rays that strike the object come from a single point, so that a given point of the image either receives light or doesn't.

If the source of light is too large to be regarded as a "point" source, then the shadow is constructed using the principle of superposition – the extended source is modeled as a large number of point sources, and the shadow produced by each of these point sources is constructed independently. This is an example of the "principle of superposition" – that is that the effect of a number of elementary sources is the sum of the effects of each one of them taken individually.

The shadow produced by two or more point sources is also handled by the principle of superposition in the same way – the light striking a screen behind the object is the sum of the contributions from all of the light sources taken individually.

The *umbra*: the portion of the shadow which is totally dark because light from all sources is absent.

The *penumbra*: the portion of the shadow which is illuminated by some of the sources and is therefore not completely dark.

In some situations (such as a transparency projector or an x-ray picture), the information content of the image is really in the shadows, and the unobstructed parts of the image are not of interest.

False shadows are often used to make a flat image appear "3-dimensional." This works because our eyes are accustomed to the idea that a shadow is produced when an object is in front of a screen.

SOLAR AND LUNAR ECLIPSES

Since the Moon orbits the earth with a period of about 29.5 days, there are occasions when the Moon comes between the Earth and the Sun or when the Earth comes between the Moon and the Sun. These are called eclipses, and they can be understood using the principles of shadows outlined above. (The orbit of the Moon is not exactly in the same plane as the orbit of the Earth around the Sun, so that an eclipse does not happen *every* 29.5 days.)

Although the Moon is much smaller than the Sun, it is much closer to the Earth, so that its shadow can totally block the light of the Sun if the alignment is exactly correct. This is a "total" eclipse; it is quite rare, since the alignment is usually not exact. The more usual situation is a "partial" eclipse in which only part of the Sun is blocked. Even during a total eclipse, the umbra of the Moon does not completely cover the Earth, so that only some places on the Earth see the eclipse as total. Other locations see only a partial eclipse, and yet other locations see nothing at all, since they are neither in the umbra nor in the penumbra.

A lunar eclipse occurs when the Earth comes between the Sun and the Moon. As above, the eclipse may be total or partial, depending on the exact alignment. Since the Moon shines by reflecting light from the Sun, the moon almost disappears during an eclipse. However, even in a total eclipse, it is often possible to see a very faint image of the Moon because some light from the Sun is deflected by the atmosphere and still reaches the Moon.

THE PINHOLE CAMERA

THE RELATIONSHIP BETWEEN AN OBJECT AND ITS IMAGE

An object can be anything that we observe using an optical system. If it is a physical object, then we imagine that light rays are emitted in all directions from every point on the object. The intensity (and the colour) of these rays may vary from point to point. An object may also be virtual. A virtual object is usually the image of a physical object formed by an optical system. An example would be the person that looks back at us when we look in a mirror.

If an optical system forms an image, that means that there is a one-to-one relationship between a point of the object and a point of the image. That is, a point of the image receives light from only one point of the object. The object and the image need not be the same size or have the same orientation. The requirement is only that any given point of the image receive light from only one point of the object.

This one-to-one relationship is usually not realised when light from an object is simply incident on a screen or other target. A single point on a screen usually receives light from many different points on an object, so that there is no unique one-to-one relationship between object and the light that strikes a point on the screen.

We see distinct objects because the lens at the front of our eyes produces an image of the scene on the retina – the light-sensitive "screen" at the back of the eye. This image is really our only connection to the real world, and we can be fooled in interpreting this image if the assumptions that we unconsciously make in evaluating it are not true. This is the basis for "optical illusions," for our ability to see three-dimensional effects in two-dimensional pictures, etc.

THE PINHOLE CAMERA

The pinhole produces the one-to-one relationship between the object and the image because of its small size. The image is dim because most rays from the object can't contribute to the image The image is inverted (both with respect to up/down and right/left) – a feature of many imaging systems

Advantages: cheap and simple; can be used for any wavelength; objects at any distance produce clear images (great "depth of field") Pinhole approximation gets better as hole gets smaller Image also gets dimmer – fewer rays can get through Watch out for λ/D if pinhole gets too small –

requirement for geometrical optics may not be true if pinhole is small enough

There is a simple relationship between the size of the object and the size of the image. This relationship can be seen from the geometrical picture shown below. The object is an arrow on the left that points upward as shown. The image is constructed by drawing straight lines from points on the object to points on the imaging screen, which is shown to the right of the pinhole. Because the pinhole is so small, only one ray from each point on the object reaches any point in the image plane.

The two triangles formed by the object and its two rays and the image and its two rays are similar, and the sides are therefore proportional. If the lengths of the object and image are L_o and L_i respectively, and if the corresponding distances between the pinhole and the object and image are D_o and D_i respectively, then using the properties of similar triangles: Magnification= $L_i/L_o = D_i/D_o$

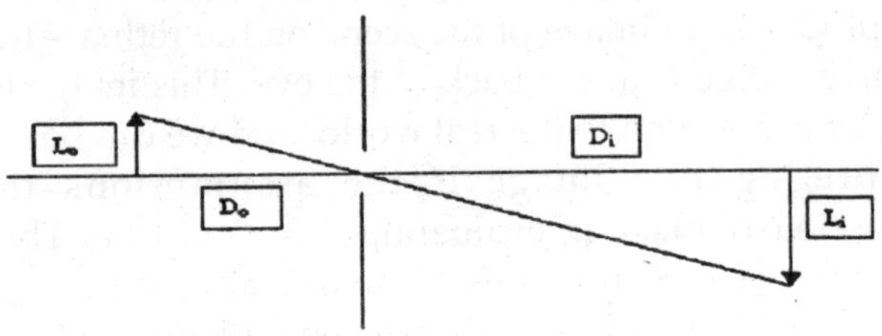

REFLECTIONS, WAVES ON A ROPE

Model a rope as a series of small masses connected by rubber bands. When one of the masses is displaced from its equilibrium position, it exerts forces on its neighbours. These forces cause the adjacent masses to move and also act to bring the initial mass back to its equilibrium position. If all of the masses and rubber bands are the same then the displacement propagates from one mass to the next one. In the absence of friction or other loss mechanisms, the resulting wave propagates without any decrease in its amplitude – each mass transmits its energy to the next mass in the chain through the rubber-band link between them.

The propagation of the wave consists of an exchange between two types of energy: the elastic potential energy that is stored in the stretched rubber bands is converted to kinetic energy of the moving masses, and this kinetic energy in turn is converted back into elastic potential energyy when the next rubber band is stretched. This is analogous to a light wave which propagates by an exchange between electric and magnetic potential energies.

The velocity of propagation in this case depends on two quantities:

1. The stiffness of the rubber bands, which determines the magnitude of the force that is transmitted to an adjacent mass for a given displacement,
2. The size of each mass, which determines how quickly the mass will respond to the transmitted force. The velocity of propagation varies directly with the stiffness of the rubber bands and inversely with the mass at each junction. (A more detailed analysis shows that the velocity of propagation varies as the square root of the ratio of these two quantities.)

When the disturbance reaches a fixed object (or a mass that is much greater than any of the others), the same force produces a much smaller displacement (possibly no displacement at all if the object is fixed), so that the restoring force is greater than at any other point on the rope. The result of the larger restoring force is to generate a reflected wave with a displacement that has the opposite sign. Using the previous paragraph, a larger mass implies a smaller propagation velocity.

When the disturbance reaches the free end of the rope (or a mass that is much smaller than any of the others), the same force produces a much larger displacement than at other points so that a reflected wave of the same sign is generated.

In the more general case, a reflection is generated whenever a wave reaches a point where its velocity of propagation changes. The reflected wave is reversed when the velocity decreases at the boundary and is not reversed when the velocity increases. Using the various definitions of the phase of a wave, the reversal at the reflection can be thought of as a phase change of one-half of a wavelength or as 180°ð in a sine-wave cycle or as a time delay of one-half of the period of the wave.

Our eyes and many other optical detectors are sensitive only to the amplitude of the light wave (or to the intensity, which is proportional to the square of the amplitude) and not to its phase, so that these reversals produce no visible change in the light and can be ignored. However, this is not always the case, and the phase-reversal on reflection plays an important role in interference between two light waves.

REFLECTIONS, RADAR AND SONAR

If the velocity of propagation of a wave is known, the time it takes the reflected wave to return to the source is a measure of the distance between the source and the object that produced the reflection. This is the basis for radar, sonar and time domain reflectometry (TDR), all of which measure the distance between a transmitter and a reflector using the time it takes the echo of a pulse to return to the sender.

In order for radar and sonar to be useful, the system must operate in the limit where geometrical optics is valid. In other words, these methods will only "see" an object if its size is considerably larger than the wavelength that is used, so that the ratio of λ/D is less than 1. For example, if a bat is to detect an insect whose size is about 6 millimeters in diameter (about 0.25 inches), the bat must use a sound wave whose wavelength is significantly smaller than this value. The velocity of sound in air is about 335 m/s. If we take a wavelength of 3 mm as just barely satisfying the criterion of being not larger than the object that is to be detected, the frequency of the sound wave must be at least $\nu = c/\lambda = 335/3 \times 10^{-3} \sim 112$ kHz

This is in fact about the frequency that bats use. It is much higher than we can hear. Shorter wavelengths (that is, higher frequencies) allow the bat to detect smaller objects or to see the structure of a larger object.

Note that we use the symbol "c" for the velocity of sound in air (rather than "v") to avoid confusing it with the frequency ν.

Time domain reflectometry is often used to detect the location of a broken power cable in a remote region. When a pulse of voltage is applied to one end of the cable, the broken end of the cable causes a reflection to be sent back to the transmitter. The time it takes the pulse to return gives the approximate location of the break. The same technique is used to detect the location of a break in an underground telephone cable.

Unlike our eyes or our cameras, which are simply receivers that detect objects using reflected light that is produced by some external source, radar and sonar systems also must include an active transmitter. They are therefore both more complex and more versatile – they can detect objects even in the absence of ambient light, but they must include a separate transmitter to make this possible.

The range of radar and sonar systems is limited by two factors. The maximum range is determined by the amount of power that the transmitter can produce and by the sensitivity of the receiver that is used to detect the echo. The minimum range is given by the timing resolution of the receiver – that is, by the ability of the receiver to distinguish between the original transmitted pulse and the (usually much weaker) reflected echo. When the range to the target is too small, the echo can arrive so quickly that the transmitted pulse has not yet died away. The goals to be able to detect objects that are very far *or* very close to the transmitter using a single system conflict with each other to some extent. The maximum range of an echo system is increased by increasing the transmitted power, but this increase will saturate the receiver for a longer period of time and make it more difficult to detect an object that is very nearby. Radar systems have clever pulse shapes (called chirps) that attempt to minimize this conflict. Bats use the same techniques.

REFLECTIONS,METALS AND DIELECTRICS

The simple model of a metal is a regular, rigid arrangement of positively charged ions and a cloud of electrons that are relatively free to move inside the material. The free electrons respond to applied electric fields. Since the electrons are approximately free particles, they arrange themselves on the surface of the metal so as to cancel the electric field in the interior. For example, an external positive charge attracts electrons to the surface of the metal. This attraction continues to draw more electrons to the surface until the repulsion between these electrons at the surface exactly cancels the attraction of the external positive charge.

Since the electrons in a metal are free particles and the electric field does not penetrate into the interior of the material, very little of the incident energy is absorbed, and most metals are good reflectors. Ionized gases, which are also made up of positive ions and relatively free electrons behave very much like metals in this regard, and tend to be good reflectors as well. Atoms near the top of the atmosphere are ionized by ultraviolet radiation from the sun, and this layer of ionization is a good reflector of radio waves.

Most liquids do not contain free electrons, but they may contain positive and negatively charged ions, which can respond to an electric field. However, the ions are usually not as free as electrons are in a metal, and most liquids are more like dielectrics even if they have lots of charged ions.

The approximation of completely free electrons and passive, fixed ions that is used to model a metal breaks down for two reasons as the incident frequency is raised. The electrons are not really free particles and cannot respond infinitely rapidly to changes in the electric field. This means that they become unable to follow the field as its rate of change increases. In addition, the positively charged ions will interact with the incident field as well, especially when specific frequencies (which are characteristic of the particular ion are incident). Thus metals tend to be reflectors at low frequencies, but absorbers as the frequency is raised. Some metals (gold, for example), absorb the blue light at the high-frequency end of the visible spectrum and appear yellow as a result. This effect is even more pronounced in copper, which absorbs most of the visible spectrum. Since it reflects only the lowest-frequency portion of the visible spectrum, it appears reddish-orange as a result.

The same thing is true for ionized gases – they tend to change from reflectors to absorbers as the frequency of the incident radiation is raised. If the density of the gas is not too great, there may not be enough atoms to absorb all of the incident radiation, and the gas may become approximately transparent as a result. Thus the ionosphere reflects radio waves but transmits signals in the visible portion of the spectrum. It becomes strongly absorbing for ultraviolet frequencies (above the violet end of the visible spectrum) because these frequencies are strongly absorbed by oxygen and other atmospheric gases.

The simple model of a dielectric is a substance which may have the same regular structure as a metal, but with very few free electrons. The details of the absorption and reflection depend on the details of the material, and it is usually not possible to give a general rule as it was with metals. However, dielectrics generally transmit and reflect the incident energy with coefficients that are specific to the material and are usually functions of the incident frequency. This dependence on frequency is usually much more complicated than the relatively simple variation that characterizes metals and ionized gases.

MIRRORS

Since metals make the best general-purpose reflectors, they are usually used to make mirrors. Silver has a very high reflectivity across the entire visible spectrum and is usually the first choice for a mirror; aluminum is almost as good. Both of these metals interact with the oxygen and with the other components in the atmosphere to form compounds (mainly oxides and sulfides) that are not metallic and do not reflect nearly as well as the pure metal itself. The metallic reflecting surface is therefore protected in some way. Silver mirrors usually have a pane of glass in front of them; aluminum mirrors are often coated with a thin layer of aluminum oxide. In both cases, the protective layer degrades the reflectivity somewhat but protects the underlying metallic surface from oxidation.

The simple model of a metal as a perfect reflector breaks down if the metallic layer is very thin (that is, only a few atoms thick), so that there is not much difference between the surface of the material, where the external electric field is concentrated and the interior, which we modeled as a field-free region. These mirrors are often called "half-silvered" because the thickness of the layer is adjusted so that about 50per cent of the incident light is reflected. Since very little light is absorbed, the other 50per cent must be transmitted, and these mirrors are often used as "one-way" mirrors.

If a one-way mirror is placed between two rooms that have very different levels of illumination, a person standing in the bright room cannot see through the mirror because the 50per cent of the light that is reflected on the bright side is greater than the 50per cent of the much lower level of illumination on the dim side. A person on the dim side, however, can easily see into the bright room because 50per cent of the light from the bright side is transmitted.

The same effect can be produced with ordinary window glass, even though it is not a 50/50 reflector. Window glass typically reflects about 4per cent of the light that strikes it, but even this small reflection may be significantly larger than the 96per cent of the light transmitted from the other side. Thus ordinary windows act as mirrors for people inside a house at night, while outsiders can easily see in. The reverse is true during the day, when the outside is much brighter than the inside. People inside the house can easily see out, whereas those outside cannot see in because they cannot detect the light transmitted from the interior in the presence of the 4per cent of the light outside the house that is reflected from the windows. The windows therefore look like mirrors to those outside.

WAVEFRONTS AND HUYGENS' PRINCIPLE

A wavefront is a surface over which an optical wave has a constant phase. For example, a wavefront could be the surface over which the wave has a maximum (the crest of a water wave, for example) or a minimum (the trough of the same wave) value. The shape of a wavefront is usually determined by the geometry of the source. A point source has wavefronts that are spheres whose centres are at the point source. A fluorescent tube would have wavefronts that are cylinders concentric with the tube itself. A very large sheet of material that is uniformly illuminated would generate wavefronts that are plane waves parallel to the sheet.

The direction of propagation of the wave is always perpendicular to the surface of the wavefront at each point. Thus, the wavefronts of a point source are spheres and the wave propagates radially outward – the radius of a sphere is perpendicular to its circumference at each point. The same thing is true of the radius of the cylindrical wavefronts that would be generated by a fluorescent tube.

Although the wave fronts produced by a point source are always concentric spheres in principle, when the source is very far away the radii of the spheres are so large that they look like plane waves to an observer. (Just as the Earth looks flat when viewed from a point near its surface.)

Huygens' principle (Christaan Huygens, 1629-1695, published about 1690) describes how a wavefront moves in space. According to this principle, we imagine that each point on the wavefront acts as a point source that emits spherical wavelets. These wavelets travel with the velocity of light in the medium. At any later time, the total wavefront is the envelope that encloses all of these wavelets. That is, the tangent line that joins the front surface of each one of them. A simple example of how a plane wavefront moves is shown below:

The same construction is used for a wavefront of any other shape.When a wave travels in a single medium at a constant speed, the Huygen's construction preserves the general form of the wavefront. That is, spheres propagate and become larger spheres, cylinders become larger cylinders, etc.

If a portion of the wavefront enters a different medium (enters glass from air, for example), then the wavelets generated by each portion of the wavefront travel with the velocity that is appropriate for the medium that the wavefront is in. That is, the wavelets in the medium where the speed of light is less will have smaller radii than the wavelets in the original medium.

Although Huygens' principle was initially stated without any proof, a slightly modified form of it was later derived by Fresnel from the mathematical theory of waves. Note that Huygens was a contemporary of Newton, and that it would probably have been much more difficult to publish his theory if he had lived in England, where disagreeing with Newton was not an easy or popular position to take.

Mirrors, The Law of Reflection

When a light wave is incident on the boundary between two media, the "angle of incidence" is the angle between the incident rays and the normal to the surface at that point. The reflected ray emerges at the same angle with respect to the normal (the perpendicular to the surface at that point). The incident ray, the reflected ray and the normal all lie in the same plane. This result can be derived from Huygens' principle using a simple geometrical construction of the incident wavefront.

Since the wave is incident on the boundary between the two media at some angle other than 0 degrees or 90 degrees, one part of the wavefront reaches the boundary before the rest of the wavefront. The reflected wavelets from that part start radiating backwards while the remainder of the wavefront is still moving towards the boundary. The reflected wavefront can be completed when the last part of the incident wave reaches the boundary. A simple geometrical construction shows that applying Huygens' principle to this case results in the angle of incidence being equal to the angle of reflection.

If the reflecting surface is plane and smooth then a parallel beam of light produces a reflection whose rays are also parallel. This is "specular" reflection. If the surface is curved or not smooth, then the direction of the reflected beams will vary from point to point, and a parallel beam of light will produce a more complex reflection. In the limit of a reflecting surface that is very rough and irregular, the reflected light goes in all directions with more or less the same intensity. This is called "diffuse" reflection, in which the initial directivity of a parallel beam is completely lost. Note that this diffusion is due to the fact that the angle of incidence varies rapidly from point to point (even though the incident beam is composed of parallel rays) because the surface roughness means that the normal to the surface varies from point to point.

The law of reflection relating the angle of incidence to the angle of reflection is true for all materials, but the *amplitude* of the light reflected in a particular direction depends on the angle of incidence, on the type of material (metal or dielectric) and on the polarization of the incident radiation. This variation turns out to be much more important for reflection from dielectrics than for reflection from metals.

For most dielectrics, the amplitude of the reflected light increases as the angle of incidence increases, and most dielectrics become almost 100per cent reflectors at large angles of incidence (approaching 90°ð). Furthermore, the amplitude of the reflected light depends on the polarization of the incident beam, and there are certain angles where the reflected amplitude goes to 0 for one polarization so that only other polarizations are reflected.

These principles are illustrated by reflection from a roadway. When the road is dry, its surface is very rough, and the reflection is diffuse. Incident light from any direction is reflected in all other directions with essentially equal amplitude, so that light reflected from the road reaches our eyes and we can see it at night. When the road is wet, the surface becomes much more regular, and the reflection tends to be specular rather than diffuse. Light from an overhead street light tends to be reflected back upward and not into our eyes (the angle of incidence is quite small and therefore the angle of reflection is small as well). Light from the headlights of the car tends to be reflected forward – both the angle of incidence and the angle of reflection are large in this case. Very little light is reflected back to our eyes both because reflection is specular. Since the reflection coefficient tends to approach 100per cent at large angles of incidence, oncoming traffic sees a large amount of light that is reflected from the roadway at large angles of incidence. Most of the light from the headlights is of almost no use in seeing the road and simply causes glare for the oncoming traffic.

Refraction and Snell's Law

When light strikes the boundary between two different media, a portion is reflected according to the laws of reflection, and another portion enters the medium. The ray which crosses the boundary and enters the second medium has been *refracted*, and the process at the boundary is called *refraction*.

The velocity of light depends on the properties of the medium it is traveling in, and especially on the density of the material. The velocity of light decreases when light enters a denser medium. The ratio of the speed of light in vacuum to the speed of light in any material is called the index of refraction of that material. If c is the speed of light in vacuum and v is the speed in some material, the index of refraction is $n=c/v$

Since v is always less than c, the index of refraction is always greater than 1. Note that the index of refraction is a ratio of two speeds and therefore has no units. Another way of expressing the same relationship is:$v=c/n$ That is, the speed of light in any medium is the speed in vacuum divided by the index of refraction.

The index of refraction depends on the detailed properties of the material and may also vary with external parameters such as ambient temperature, etc. However, we will generally ignore these variations and assume that the index of refraction is a constant that is a characteristic of the material. In addition, although the index of refraction of the atmosphere is about 1.0003, we will often take this index to be exactly 1, so that the speed of light in air is equal to its vacuum value. Using Huygens' principle, it is easy to show that when the velocity of light decreases at the boundary then the refracted light is bent towards the normal. That is, the angle between the refracted ray and the normal is smaller than the angle between the incident ray and the normal. Conversely, when light crosses a boundary into a medium where its velocity is greater, the angle of the refracted ray with respect to the normal is larger than the angle of incidence. This relationship is called *Snell's Law*; the mathematical form of the law (which we will not discuss) relates the incident angle, the refracted angle and the indices of refraction of the two media at the boundary. (The relationship actually involves the sines of the angles rather than the angles themselves. Since light that leaves a denser medium and enters a less dense one (glass to air, for example) is bent *away* from the normal, the angle of refraction will always be greater than the angle of incidence in this case. Therefore, the angle of refraction can exceed 90 degrees if the angle of incidence is large enough. There is no refracted wave at this angle, which is called the *critical angle* – the light is only reflected. This is called *total internal reflection*. There is also no refracted wave for all angles of incidence larger than the critical angle. This effect is the basis for optical fibers, which effectively "trap" light inside the fiber because of total internal reflection.

There are a number of optical effects that are caused by the variation of the index of refraction. A common effect is a mirage, which is caused by the variation of the index of refraction of air with temperature. The air near the surface of a roadway is often much hotter than the air at higher elevations, and its density and therefore its index of refraction are smaller. Light incident on this air from above is therefore bent away from the normal. The light that finally strikes the road is almost perfectly reflected, since the angle of incidence is close to 90 degrees, and these reflected rays will appear to come from a virtual object that is below the surface of the roadway. Since this effect is common over water, our eyes often assume that the hot roadway is actually a body of water.

Dispersion, Prisms and Rainbows

The index of refraction of all materials varies with the frequency or wavelength of the incident light. This effect is called *dispersion.* Using quantum mechanics and electromagnetic theory, it is possible to show that this variation is a direct consequence of the fact that atoms and molecules absorb and emit only discrete frequencies or wavelengths.

In general, the index of refraction for most materials in the visible portion of the spectrum increases with increasing frequency. Thus the index of refraction for blue light is larger than for red light. This is not a fundamental principle of physics – it just happens to turn out that way. In fact, there are a number of materials where the inverse is true – it just happens that these situations are not usually observed in everyday experience.

Since the speed of light is equal to the vacuum speed divided by the index of refraction, the increase in the index from red to blue means that blue light travels more slowly than red light in most materials. In addition, the variation of the index with frequency or wavelength means that blue light is bent more than red light when a light beam crosses the boundary between two media. This was first observed by Newton.Angular dispersion, in which different wavelengths are bent through different angles at a boundary, and temporal dispersion, in which different wavelengths travel at different speeds in most media, play very important roles in everyday experience and in technology. Angular dispersion is what causes a rainbow, for example, and temporal dispersion is widely used in satellite communications systems to measure the properties of the ionosphere and the atmosphere.

Real and Virtual images; Plane Mirrors

A real image is produced on a screen (or some other detector) when all of the rays from a single point on an object strike a single point on the screen. The intensity distribution on the screen is a reproduction of the intensity distribution emitted by the object. The image may be smaller or larger; it may be erect or inverted – it is the one-to-one relationship between the image and the object that is important.

A virtual image is produced when rays of light reach our eyes that appear to come from a real object, but there is in fact no object at the apparent source of the light. The most common example is when light from an object strikes a simple plane mirror. The reflected rays appear to come from an identical object that is located behind the mirror. The spatial distribution of rays is completely consistent with what we would see if there really was an object behind the mirror. Note that we cannot actually place a screen at the point where the image appears to be.

The law of reflection – that the angle of incidence is equal to the angle of reflection, can be used to construct the image from any object placed in front of a mirror. This construction is relatively simple when the mirror is a single flat surface. A ray from object point O that strikes the mirror results in a reflected ray that appears to be coming from image point I. This construction is the same for every ray from O that hits the mirror anywhere, so that there is a one-to-way relationship between the points O and I. However, point I is behind the mirror, and is therefore a virtual image. However, the rays reflected by the mirror are completely consistent with those that would be produced if there were a physical object at point I.

This construction can be completed for any point on the object side of the mirror, so that the image space is an exact reproduction of the points on the object side of the mirror. The reflected rays preserve the size and shape of the object, so that the magnification (the ratio of the size of the image to the size of the object) is always 1. The image point I always appears to be the same distance behind the mirror as the object point O is in front of it. The mirror has therefore reversed the coordinate axis perpendicular to the plane of the mirror – object points that are further to the left of the mirror in the figure below have image points that appear to be further to the right in the virtual image.

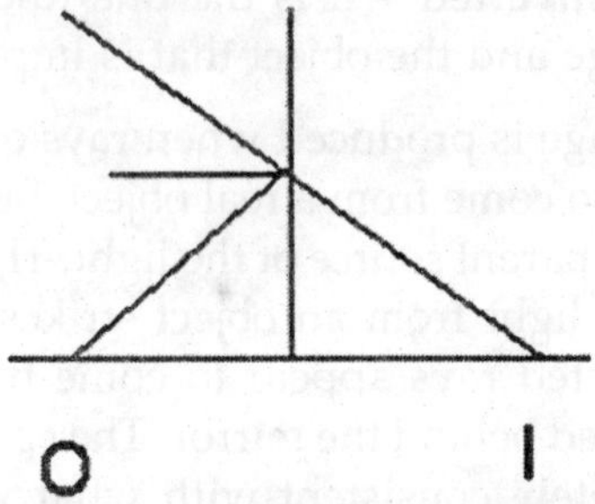

If the geometrical construction above is repeated for a point just above point O, the resulting image point will be just above point I. Thus the mirror preserves the vertical relationship between objects.

Also note from the figure that it is possible to see the image of an object O even if the mirror does not extend that far, since the reflected ray only uses a higher point on the mirror. For example, if the object O is your shoe, then you can see the reflection of your shoes with a mirror that is only one-half of your height.Finally, these constructions are exact – the image is a perfect replica of the object (within the limits imposed by geometrical optics). This is the last time in our study of mirrors and lenses that we will be able to say that.

In addition to the single-mirror geometry shown above, there are a number of useful configurations of two (or more) mirrors. Two of these configurations are shown in the following diagrams:

1. Two parallel mirrors (or any two parallel surfaces) produce an output beam that is offset in space but parallel to the input:

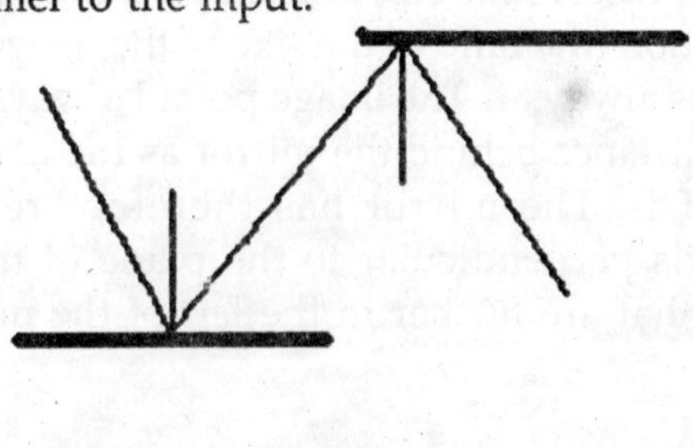

2. Two mirrors at 90 degrees (a "corner cube") reflect a light beam back along its direction of incidence, no matter what the direction was.

These constructions are special cases of two mirrors that are tilted with respect to each other. The output beam is offset by 360 degrees minus twice the tilt angle.

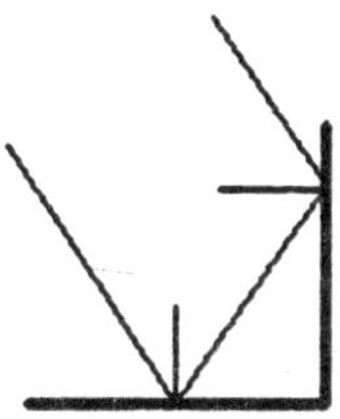

In general, if a mirror is tipped by some angle, then both the angle of incidence and the angle of reflection change by this tip angle. The angle between the incident and reflected beams therefore changes by twice the tilt angle.

FIELD OF VIEW

The field of view is the size of the largest object that can be seen through an optical system by a detector that is located at a fixed position. The size of the field can be expressed in units of length (such as meters), although it is more convenient to use an angular measure (e.g., degrees) in some situations.

For example, if you are looking at your own image in a plane mirror, then both the object (your shoes, for example) and the detector (your eye) are the same distance from the mirror. The figure below shows that a ray of light can travel from your shoes ("S" in the figure) to your eye ("E" in the figure) by striking a mirror ("M" in the figure) that is considerably smaller than your own height. In fact, in order to see your entire body at once, the mirror need be only one-half as tall as you are. You can verify this by noting the symmetry of the figure or by using simple plane geometry.

Although the principles are unchanged, the detailed situation is somewhat more complicated when the object and the detector are not at the same distance from the mirror. For example, consider the field of view of the side mirror in a car. Your eye (shown as "E") can see any object whose rays fall within the area bounded by the two rays that strike the edges of the mirror (shown as "M") as shown. Note that these two extreme rays have different angles of incidence and therefore are not parallel to each other either before they strike the mirror or afterwards.

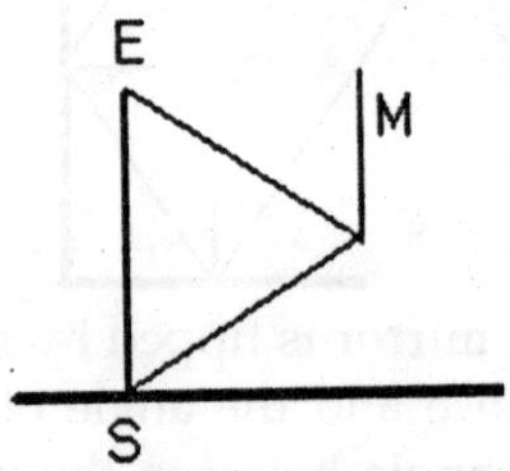

Since the two extreme rays are not parallel to each other, the field of view is a pie-shaped region bounded by the two rays that are shown. This region can be described either in terms of the wedge angle of the pie or in terms of its width at some distance behind the car.

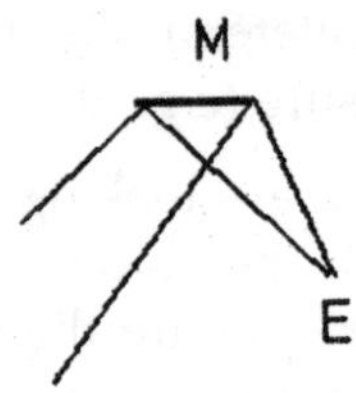

The mirror in a typical car is about 16 cm wide (about 6.5 inches), and the distance from your eye to the mirror is about 90 cm (35 inches). The field of view of this configuration is a pie-shaped region where the wedge angle of the pie is about 10 degrees. If we extend this pie-shaped region to just behind the back of the car (about 25 feet or 7.5 meters behind the driver), the width of the pie at that point is about 1.3 meters (about 4.5 feet). This pie-shaped region can be moved by changing the position of the mirror, which changes the angles of incidence of both rays simultaneously. However, the wedge angle of the pie and the width of the region that can be viewed remains essentially fixed, because these parameters are set by the width of the mirror and the distance between the mirror and the driver's eye, and these parameters do not change when the mirror is tilted. The region that can be viewed is less than the width of a typical lane on a roadway, so that this mirror will always have a blind spot – a region where another vehicle might be located but which cannot be seen by the driver. Rays emitted by objects in that region strike the mirror, but they cannot reach the driver's eye because there is no point on the mirror where the angle of incidence of the ray produces a reflected ray which is directed towards the driver's eye.

Spherical Mirrors

The principle that the angle of incidence is equal to the angle of reflection is also applicable to a mirror that is curved. The most common example is a spherical mirror, although other shapes (such as a mirror in the shape of a parabola) are also used. Since a line along the radius from the centre of the sphere (shown as "C" in the figure below) to the circumference is always perpendicular to the circumference, every radius line can be used to construct the normal to any point on the sphere. When an incoming ray (such as "I") is incident at some point on the sphere, it is reflected as ray R using the usual rule that the angle of incidence is equal to the angle of reflection. The reflected ray appears to come from some point that is inside the sphere as shown in the figure. This type of mirror is called *convex*, because the mirror surface is curved outward toward the object.

Although the direction of the radius vector changes from point to point, all rays that are parallel to ray I are reflected in such a way that the reflected rays appear to come from a single point. This is called the focal point ("F" in the figure) and is midway between the centre of the sphere and the circumference as shown in the figure. Unlike a plane mirror, where all of the rays really did appear to come from a single point, the situation here is only an approximation that is reasonably correct provided the rays do not strike the sphere too far away from the axis.

Since the angle of incidence and the angle of reflection are always equal, the law of reflection is symmetric, and a ray that was incident along direction R would be reflected along direction I. Thus we can formulate the 3 rules that describe the reflection of light rays from a spherical mirror:

1. A light ray that is incident along a direction parallel to the axis of the mirror is reflected so that it appears to come from the *focal point* – a point that is halfway between the centre of the mirror and the circumference along the axis.
2. A light ray that is incident along a direction that is pointed directly at the focal point is reflected parallel to the axis. This is the same as the previous rule with the roles of the incident and reflected rays interchanged.
3. A ray that is incident along the radius vector is reflected back on itself. Such a ray has an angle of incidence of 0 by definition, and its angle of reflection is therefore 0 as well. This in the only one of the 3 rules that is exactly true.

In addition to the three common rules, there is a 4th rule that is often useful. Since the mirror is always symmetric about its central axis, a ray that strikes the mirror at its centre (at any angle of incidence) is reflected symmetrically backwards by the same angle below the axis.

It is important to remember that rules 1 and 2 are approximations that are valid only when the incident rays are not too far from the axis. This is called the *paraxial ray* approximation. This approximation is an important limitation for reflections and refractions at curved surfaces that are spherical. (It is possible to remove this limitation in some cases using non-spherical shapes, but these shapes are usually too expensive to fabricate and are used only in very special situations.)

The validity of the *paraxial ray* approximation is determined by how far the incident ray is from the axis. The scale factor is the ratio of the distance of the ray from the axis divided by the focal length, and the paraxial ray approximation requires that this ratio be small compared to 1. This ratio often appears as its reciprocal: the ratio of the focal length to the size of the object (or to the diameter of the lens or the mirror), and the paraxial ray approximation implies that this reciprocal ratio be much larger than 1. This reciprocal ratio of the focal length to the size of the object or the diameter of the lens or the mirror is called the f-number, and it plays an important role in many other aspects of the formation of images by lenses and mirrors. Using this notation, the paraxial ray approximation is equivalent to the statement that this simple model of constructing the image produced by a lens or mirror is valid only when the f-number is large, and the approximation becomes increasingly poor as the f-number decreases.

The following diagram shows the rules governing the reflection of rays from a convex, that is one whose curvature is outward towards the source of the rays. Note that the fact that the angle of incidence always equals the angle of reflection implies that rays R and I can be interchanged.

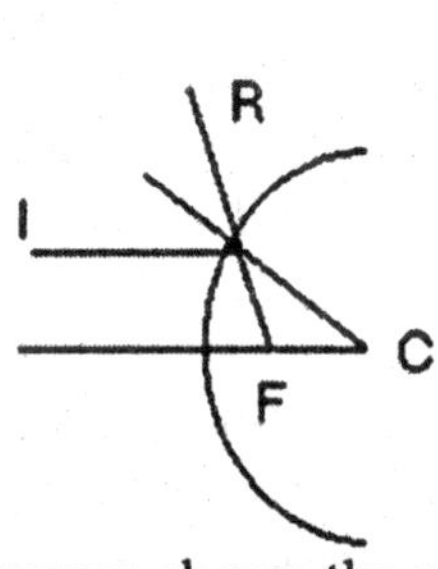

The following diagram shows the same rules applied to a concave mirror, that is one whose curvature is inward away from the source of the rays. A ray that travels parallel to the axis is reflected through the focal point, as shown. As above, the fact that the angle of incidence always equals the angle of reflection implies that this ray can also be reversed. That is, a ray striking the mirror from the focal point is reflected parallel to the axis. As above, a ray that strikes the mirror along the radius of curvature is simply reflected back along itself, since any such ray strikes the mirror with an angle of incidence of exactly 0.

This figure also shows the other two rules: a ray directed along the centre of curvature is simply reflected back on itself, since the angle of incidence and reflection are both 0. Likewise, a ray that strikes the mirror exactly on its symmetry axis is reflected below the axis by the same angle.

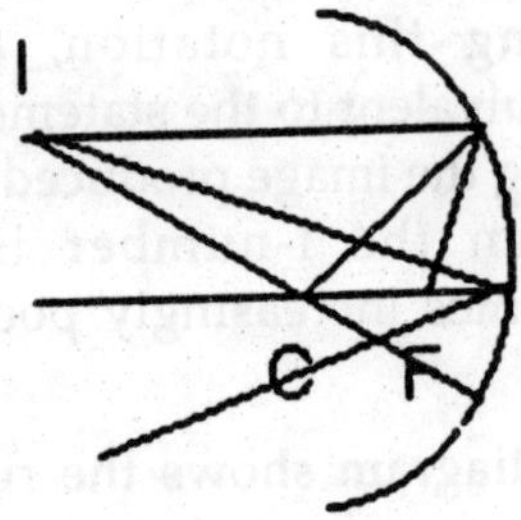

Spherical Mirrors, Image Formation.

The rules of the preceding can be used to calculate the position and size of the image produced by any spherical mirror.

1. The image produced by a convex mirror. The image is formed behind the mirror and is therefore virtual. The image is erect and is smaller than the object. The image is formed at a point behind the mirror that is almost always less than the distance from the front of the mirror to the object. The image distance varies in the same direction as the object distance – an increase in one produces a corresponding increase in the other. However, the maximum image distance is the focal distance when the object is very, very far away, and the minimum image distance is 0 when the object is right at the mirror.

The magnification is the ratio of the size of the image to the size of the object. Since the object and image vectors form the bases of two similar triangles, the magnification is the ratio of the distance of the image to the focal point divided by the distance of the object to the focal point, and this ratio is almost always less than 1. It approaches 1 as the object comes closer and closer to the mirror surface.

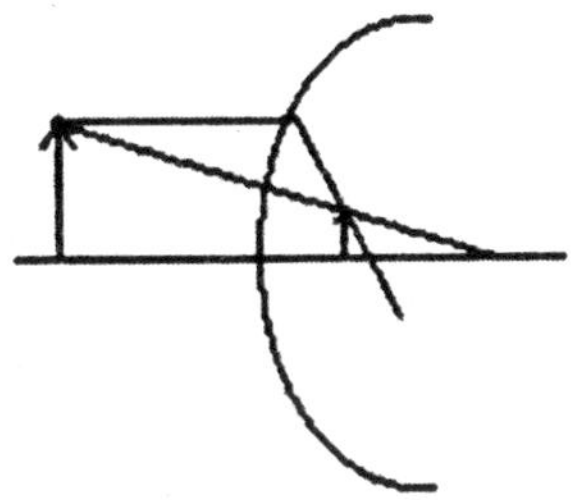

If a point source of light is very far away from a convex mirror, then, although the wavefronts are spheres, they have very large radii and so look like plane waves. The rays that strike the mirror are essentially parallel to each other in this case.

If the rays strike the mirror parallel to its axis, then all of them are reflected so that the reflected rays appear to come from the focal point. The reflected rays therefore look like a single point source of light located at the focal point.

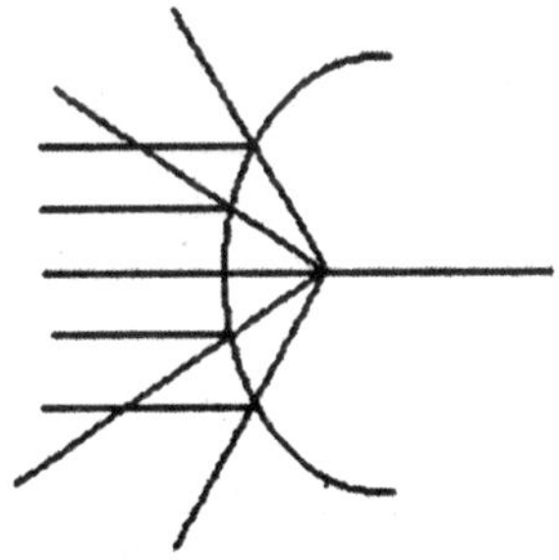

If the rays strike the mirror at an angle to the axis, then image point can be found using the usual 3 rays: one that is aimed at the centre of curvature and is reflected back on itself, one that is aimed at the focal point and is reflected parallel to the axis, and one that is directed at the centre of the mirror and is reflected symmetrically below the axis. The image point is on the focal plane, but not at the focal point. That is, it is located directly above (or below) the focal point.+

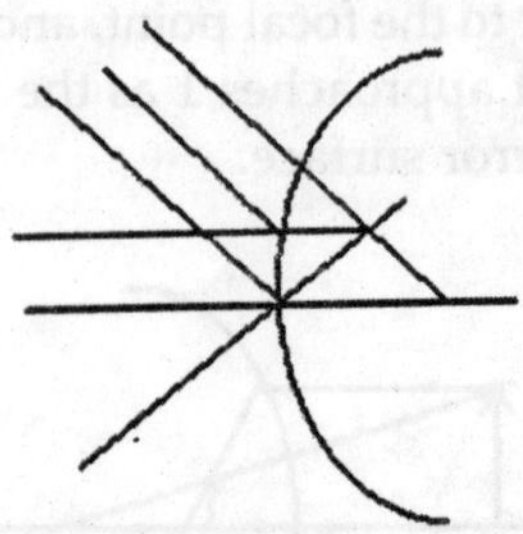

2. The image produced by a concave mirror when the object is outside of the focal point. The image is inverted and is real. The rays intersect at the image point and really do appear to diverge again from that point.

When the object is further away from the mirror than the centre of curvature (as shown in the figure), the image will be formed somewhere between the centre of curvature and the focal length. The image will always be inverted and will be smaller than the object.

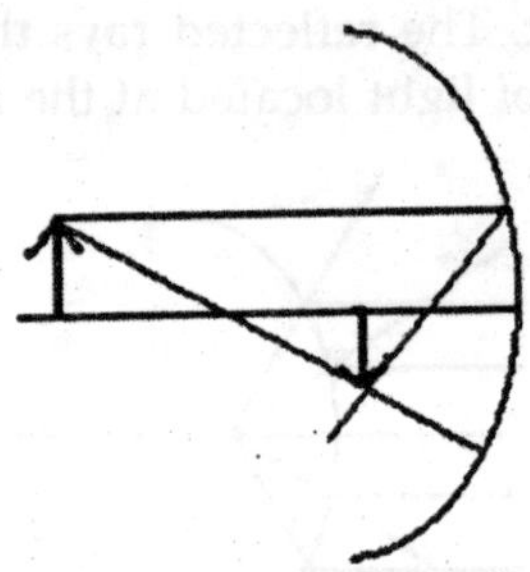

3. As the object in the diagram above moves in towards the mirror, the image moves out to meet it. The following diagram shows the object and image when the object has moved in far enough so that it is located at the centre of curvature of the mirror. No ray from the top of the object can pass through the centre of curvature in this configuration, but we can find the position of the image using other standard rays: a ray parallel to the axis and a ray that strikes the mirror exactly at its centre.

The diagram below is symmetric between the object and the image: both are located at the centre of curvature and are equal in size. The image is inverted, but it has now grown to be the same size as the object.

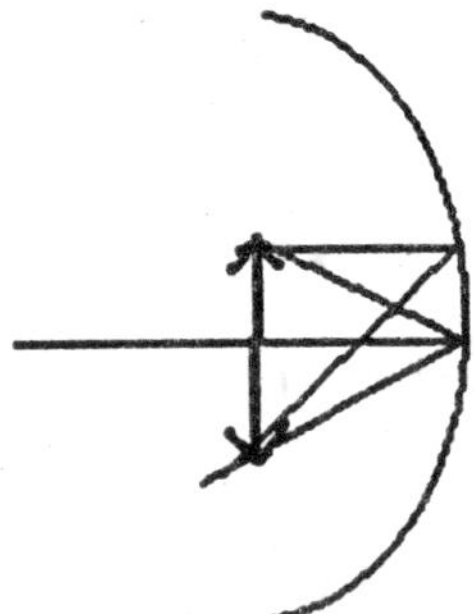

4. This process continues as the object is moved closer and closer to the mirror. The image continues to increase in size, but it remains inverted and it continues to move further and further away from the mirror. Nothing special happens as long as the object does not actually get to the focal point. The image is always inverted, and it is always larger than the object.

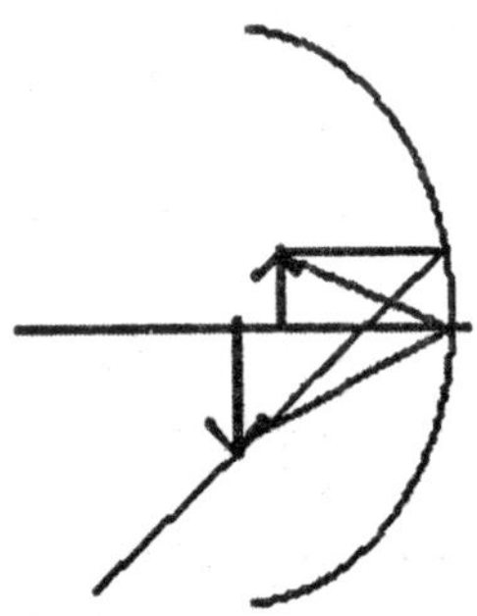

5. The image produced by a concave mirror when the object is at the focal point. The rays from the bottom of the arrow all emerge from the focal point and therefore are reflected parallel to the axis. The rays do not converge to an image point at any finite distance. We can think of this as if the image is formed at infinity, where all of the parallel lines meet again.

Rays from the top of the arrow cannot pass through the centre of curvature or the focal point. A ray that left the top of the arrow parallel to the axis would be reflected through the focal point, and a ray from the top of the arrow that was headed for the centre of the mirror would be reflected downward by the same angle. These two rays are divergent and do not converge to a single point at any finite distance. This confirms the statement that there is no image of the entire object at any finite distance – only at infinity.

The same figure shows the image produced when the object is very far away. These two cases are symmetrical rays entering parallel to the axis from infinity converge at the focal point, and rays emerging from the focal point in any direction are reflected parallel to the axis.

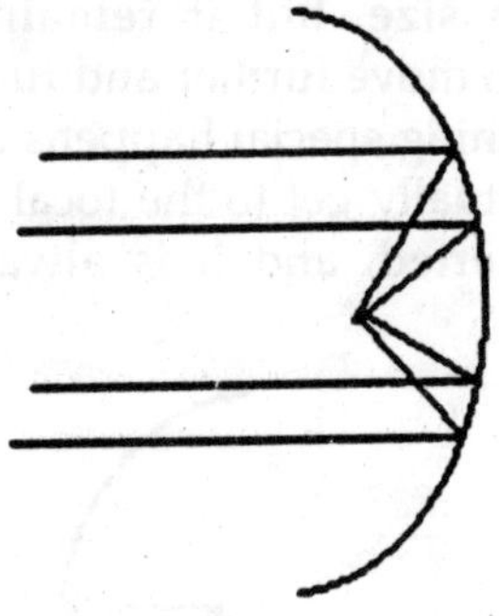

If the rays are parallel (from an object at infinity, for example), but strike the mirror at an angle to the axis, then the same thing happens here as happened above with a convex mirror. The image is formed in the focal plane, but not at the focal point. The position of the image is found using the same technique as was used above.

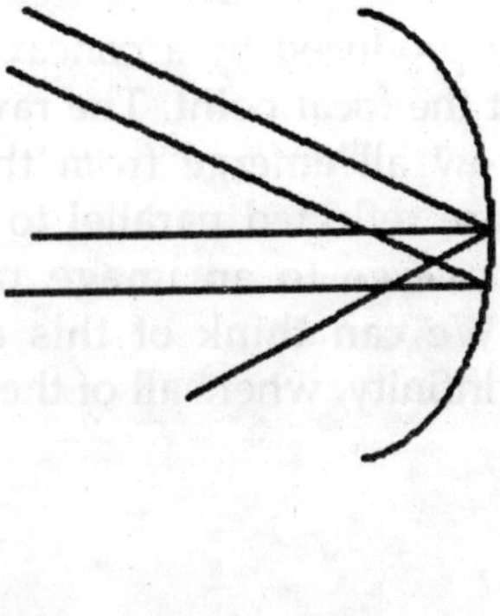

6. When the object moves inside of the focal point, then the image changes in a fundamental way. It is no longer possible for a ray from the object to actually pass through either the centre of curvature or the focal point, since the object is inside of both of these distances. However, there are still rays which leave the object along these same directions so that they might have seemed to emerge from the focal point or the centre of curvature, and these rays can be used to construct the image in this case. Likewise, a ray which leaves the object headed for the centre of the mirror is reflected symmetrically about the axis of the mirror as usual.

These rays do not converge to any finite point, and therefore there is no real image formed. However, the rays appear to be coming from a point on the right hand side of the mirror, and therefore there is a virtual image formed on the right hand side. Note that this image is not inverted as the real images above were, but is erect. Also, it is larger than the object.

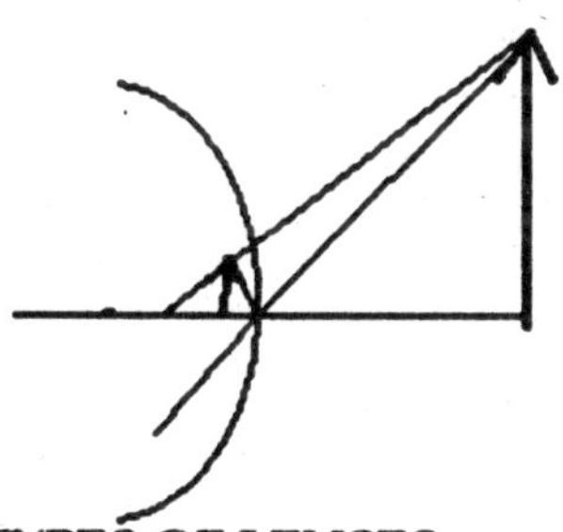

THIN LENSES, TYPES OF LENSES

In addition to using mirrors to form images, it is also possible to form images using lenses. Although many different types of lenses exist, we will consider in detail only lenses whose surfaces are spherical. In addition, the lenses we will study are made from glass or some other similar material whose index of refraction is greater than the index of refraction of the air that surrounds them. (The opposite situation – a spherical bubble of air in the centre of a liquid, for example, can be handled by the same general methods, but we will not discuss these configurations.) We will also not discuss in detail lenses that have more complicated shapes, such as those whose surfaces are in the shapes of ellipses or parabolas.

The lenses we will study fall into two broad categories: converging, or positive lenses and diverging, or negative ones. Each category has a number of different configurations, but all of the shapes in either category share a common characteristic. Converging, or positive lenses are always thicker in the middle than they are at the edges, while diverging, or negative lenses are always thinner in the middle than they are at the edges.

These are some examples of converging, or positive lenses:

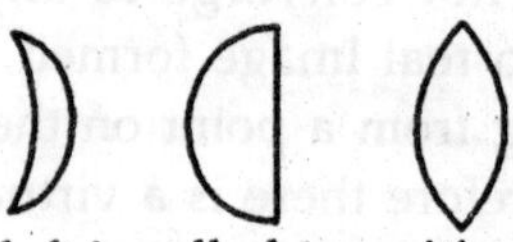

The lens on the left is called a positive meniscus (or "fish-eye") lens, the middle one is a plano-convex lens, and the one on the right is a double-convex lens. Although all of these shapes have the same basic optical properties, the first two produce better images when the field of view is very large or when the object is very far away compared to the size of the lens. The ray-tracing examples will always use the symmetric configuration on the right, but the same ray-tracing rules will work for the other two shapes as well.

Here are some examples of diverging, or negative lenses:

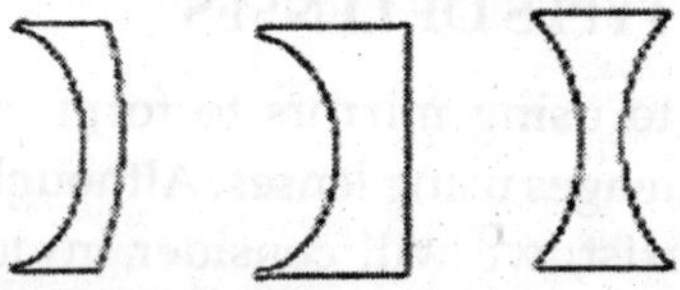

As above, the lens on the left is a negative meniscus lens, the one in the middle is a plano-concave shape and the one on the right is a double-concave shape. The choice among these shapes is governed by the same considerations as for a positive lens as described above. We will use the symmetric concave shape in the ray-tracing examples.

In each case, the properties of the lens are completely specified by its focal length. However, unlike the case of a mirror where the focal length is a simple geometrical parameter, the focal length of a lens depends both on its shape and on the index of refraction of the material it is made of. For the symmetric convex and concave shapes shown above, the approximate focal length is determined from the equation:

$$\frac{1}{f} = \frac{2(n-1)}{R}$$

where R is the radius of curvature of either surface and n is the index of refraction of the glass (or whatever material the lens is made of). The magnitude of the focal length is determined by this relationship for either a double convex or a double concave lens; the sign of the focal length is positive for a converging (convex) lens and negative for a diverging (concave) shape.

Since all lenses have light rays traveling through the lens, all lenses have two focal points: one on each side of the lens. The distance from either focal point to the centre of the lens is given by the expression above.

It is very common for relationships involving the focal length to depend not on the focal length itself but on its reciprocal. This reciprocal appears so often in optical design that it is called the "strength" or "power" of the lens and it is measured in *diopters*. The strength of a lens in diopters is the reciprocal of its focal length in *meters*. A lens with a focal length of 50 cm (0.5 m) would have a strength of 2 diopters, a focal length of 25 cm would be a strength 4 diopters, etc. Note that the focal length must be converted to meters to compute the strength of the lens. The sign of the strength of the lens in diopters is the same as the sign of the focal length in meters: positive for convex or converging lenses and negative for diverging ones.

Thin Lenses, Ray Tracing Rules

The ray-tracing rules described in this section can be used to find the position and size of an image given the focal length of the lens and the position of the object. The rules depend on the same approximations as for mirrors: the rules are valid only when the rays strike the lens in the paraxial region (i.e., close to the central axis of symmetry). In addition, the lens must be "thin," which means that its thickness must be small compared to its focal length. Many real-world lenses (such as those in cameras) do not satisfy this condition exactly, but the general principles of this section are still applicable.

Ray-tracing rules for a converging (positive) lens:

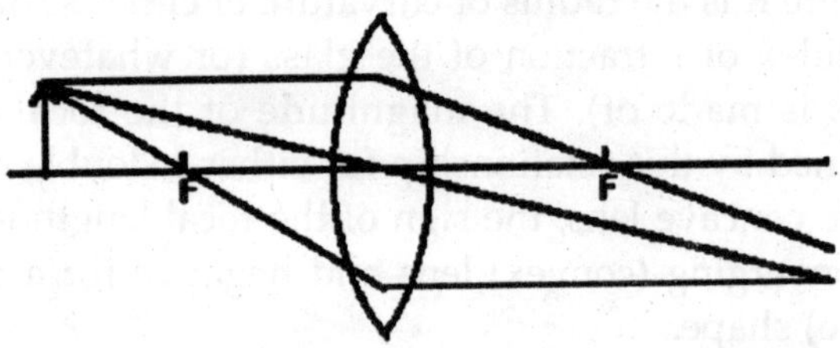

1. A ray that strikes the lens parallel to the axis is refracted through the lens and passes through the focal point on the far side of the lens. (This is often called the second focal point). See the top ray in the figure above.
2. A ray that passes through the centre of the lens emerges traveling in the same direction with no change. See the middle ray in the figure above.
3. A ray which passes through the first focal point (on the object side of the lens) is refracted so that it emerges on the other side parallel to the axis of the lens. See the bottom ray in the figure above.

Ray tracing rules for a diverging (negative) lens:

1. A ray that strikes the lens parallel to the axis is refracted through the lens so that it appears to come

from the focal point on the object side of the lens. See the top ray in the figure above.

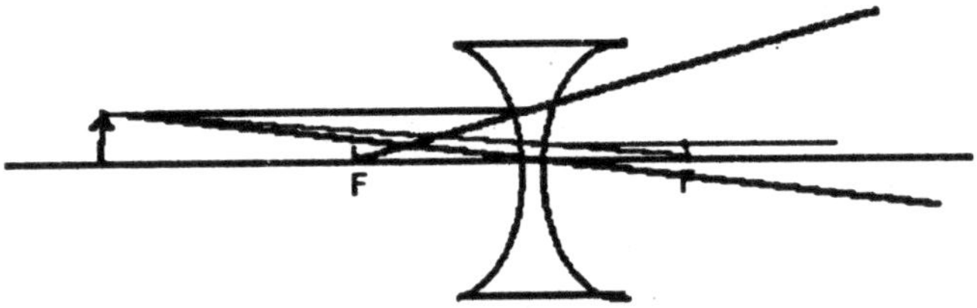

2. A ray that strikes the lens heading for the focal point on the image side of the lens is refracted through the lens so that it emerges parallel to the axis on the other side. See the middle ray in the figure above.
3. A ray that strikes the lens in the centre emerges traveling in the same direction on the image side. See the bottom ray in the figure above.

In general, only two of the three rules are needed to locate the image, and the most convenient pair of rules can be chosen in any situation.

The fundamental assumption of these rules is that *all* of the rays from any point on the object will reach the same point of the image: the rays that are chosen are only examples that are easier to draw.

Thin lenses, Images using positive lenses

1. The object is beyond the first focal point, but is at a finite distance from the lens:

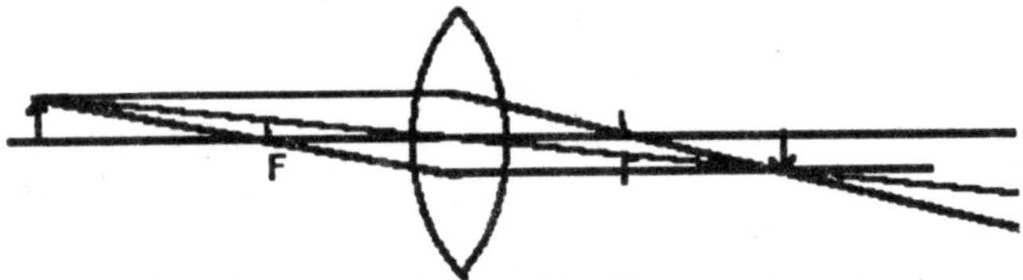

Using the three rays defined in the previous topic, we can see that the image is real, inverted and beyond the second focal point on the image side of the lens.

The image is real because the rays on the image side of the lens from any point on the object really do converge to a single point on the image. If we look into the lens from the image side, we see the rays appearing to diverge from the image that is formed.

The magnification (the ratio of the size of the image to the size of the object) depends on the exact position of the object. The magnification is less than 1 (that is, the image is smaller than the object) when the object is very far away, and it slowly increases as the object approaches the lens. It is exactly 1 when the object is twice the focal distance away from the lens, and continues to increase as the distance is reduced still further. The lens produces a magnified image when the object is located at a distance of less than twice the focal length (but greater than the focal length) from the lens.

Note that in addition to getting larger, the image moves away from the lens (to the right) as the object approaches the lens.

2. The object is at the first focal point on the left side of the lens.

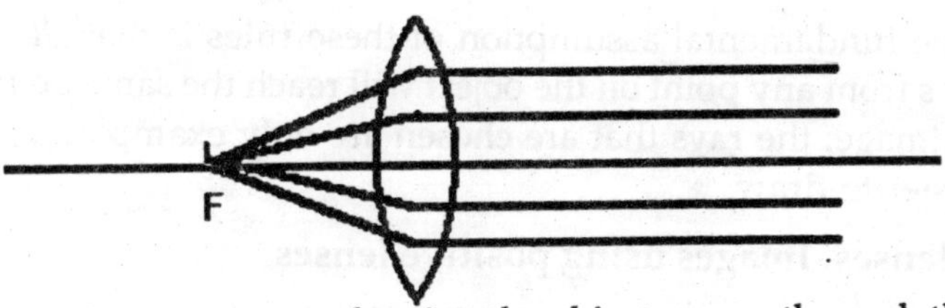

Since every ray leaving the object passes through the focal point, all of the rays emerge parallel to the axis on the right-hand side of the lens. A point object located at this position therefore produces a parallel beam of light.

Note that this configuration can be used to show that a beam of parallel light striking the lens is imaged to a point of light at the focal point. This configuration is simply the inverse of the situation that is shown and would be realised by simply interchanging the object and the image.

This configuration can be used to determine the focal length of a positive lens – it is simply the location of the image when the object is very far away. As a practical matter, an object distance of more than 10 times the focal length is sufficient for most purposes.

3. The object is inside of the first focal point on the left side of the lens.

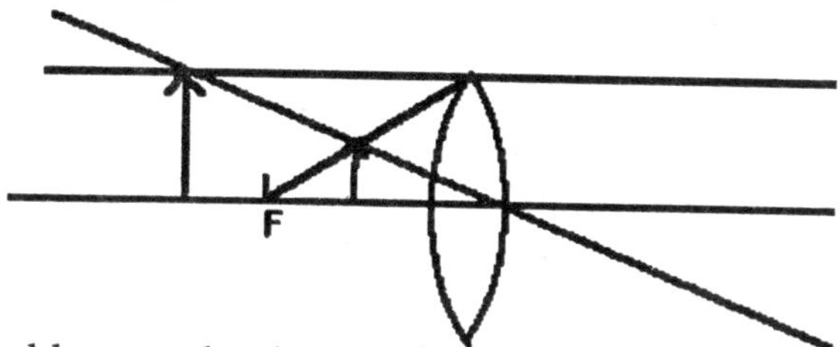

The blue ray that leaves the object comes off at the same angle as a ray that originated at the focal point, which is to the left of the object in this situation. Therefore, this blue ray is bent so that it emerges parallel to the axis on the right side of the lens.

The green ray that leaves the object heading towards the centre of the lens passes through the lens without any change in direction.

The rays do not converge to an image on the right hand side of the lens, but rather *appear* to be coming from a larger image that is behind the object. Note that the rays only appear to be coming from this image, so that this image is virtual rather than real. It is also erect – that is it has the same orientation as the object and it is magnified – its size is larger than the size of the object.

Thin Lenses, Images using Negative Lenses

The situation when a negative lens is used to form an image is simpler than for a positive lens, because the general features of the image are the same no matter where the object is located.

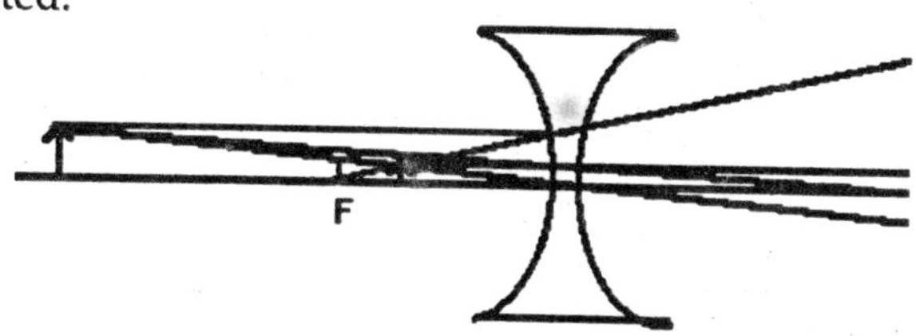

The green ray, which strikes the lens parallel to the axis emerges on the other side of the lens as if it had originated at the focal point a shown.

The blue ray, which strikes the lens at its centre, emerges on the other side without any change in its direction.The purple ray, which strikes the lens in a direction that is headed for the focal point on the opposite side of the lens, emerges on the other side parallel to the axis.These rays do not converge to an image. Rather, they appear to come from a virtual image which is smaller than the object, has the same orientation as the object, and is located between the lens and the focal point as shown in the figure.

FRESNEL LENSES

A point source of light at the focal point of a positive lens produces an image which is a parallel beam of light. This configuration is used in spotlights, lighthouses, and similar applications. In order to minimize the size of the device, the focal length of the lens must be as short as possible, since this sets the distance between the lens and the lamp. This short focal length is usually implemented in a very large lens so as to capture as much of the output of the lamp as possible. The resulting lens is large and very heavy.

Since the imaging power of a lens is produced by refraction at its edges, the centre of the lens makes almost no contribution. A Fresnel lens is designed using this idea. The interior glass of the lens is removed, and the curved surface of the lens is approximated by a series of curved segments. The effect is to preserve most of the focusing power of the lens while at the same time saving a considerable amount on its size and weight.

The design of a Fresnel lens usually starts from a plano-convex lens as shown below in the left panel. The lens is then cut into horizontal sections and each section is then aligned so that the curved edges on the right hand side of the lens are aligned. Then the extra material on the back side of the lens is removed.

The resulting lens has the same refractive performance as the original, since the shapes of both the front and the rear surfaces have not changed. However, it has much less material and is therefore much lighter.

Although it looks like the modified Fresnel should have the same performance as the original, this is not quite true. A ray through the centre of the Fresnel lens, for example, reaches a point on the image side much faster than a ray through the centre of the original lens, because a larger fraction of its path is in air. Thus a packet of rays which travel through the Fresnel lens are separated in time compared to the same packet of rays passing through the original lens. This same effect is true for rays that travel through other parts of the lens. This results in subtle changes to the image that are usually not important for spotlights and lighthouses, but can be significant whenever the phase of the light is important.

To build a spotlight or a lighthouse, a Fresnel lens is usually combined with a concave mirror. The concave mirror is placed behind the lamp with the lamp at its centre of curvature. Every ray that leaves the lamp traveling backwards strikes the mirror perpendicular to its surface in this case and is therefore reflected back on itself. These reflected rays therefore travel in the same directions as the rays that emerge from the front of the lamp. The lamp is placed at the focal point of the Fresnel lens, so that all of these rays emerge from the lens in a parallel beam. This configuration uses most of the light that is emitted by the lamp, although the rays that do not strike either the mirror or the lens are wasted.

Chapter 7

Compound Lenses

The ray-tracing methods that we have developed can also be applied to a number of lenses that are close together. These combinations are called compound lenses, and are very commonly found in cameras and other optical instruments.

In order to apply the standard ray-tracing method to the combination, we start by finding the image that would be produced by the first lens acting alone. For example, consider the compound lens shown below.

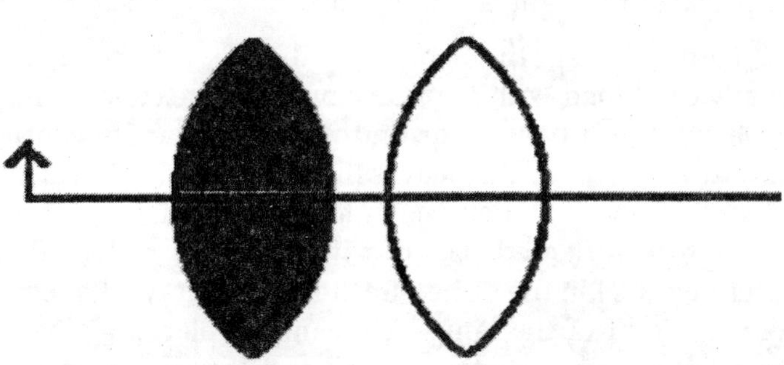

In order to find the location of the image produced by the combination of the two lenses, start by using the usual ray-tracing rules to find the image produced by the first lens. The image is located using the usual two rays: one that strikes the lens parallel to the axis and emerges directed towards the focal point and the second which strikes the lens at its centre and continues without any change in direction.

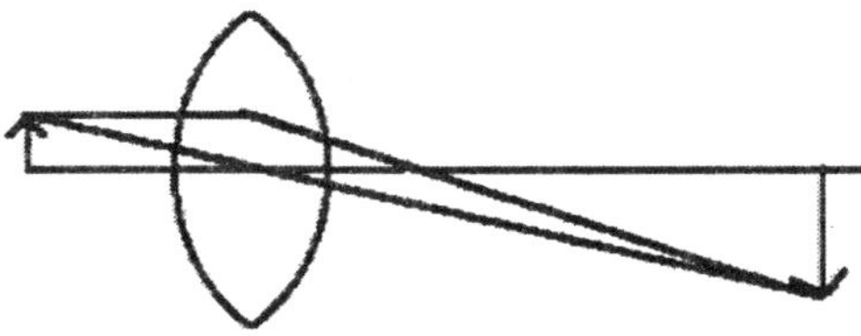

Once the image has been found, we can add any number of additional rays from the object to the image without the need for any rules, because every ray that leaves a point on the object and strikes the lens is refracted so that it converges on the corresponding point of the image. Although we can choose any additional rays, it is most convenient to choose rays that will make the next step easiest. Thus we choose rays in this step that will pass through the centre of the second lens and will strike the second lens parallel to the axis. Note the two additional rays have been chosen with this principle in mind.

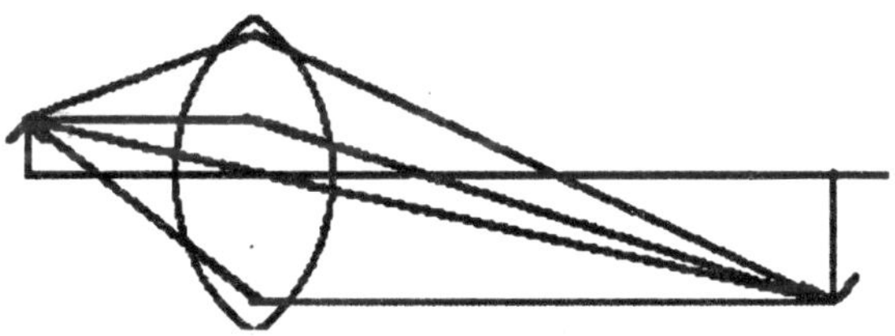

We then apply these rays to the second lens and use the usual rules to compute the location of the second image. This job was made easy by choosing additional rays which would satisfy the usual rules with respect to the second lens. That is, one of our additional rays passes through the centre of the second lens and therefore emerges unchanged and the second ray strikes the second lens parallel to the axis and therefore emerges directed towards the focal point of the second lens as shown.

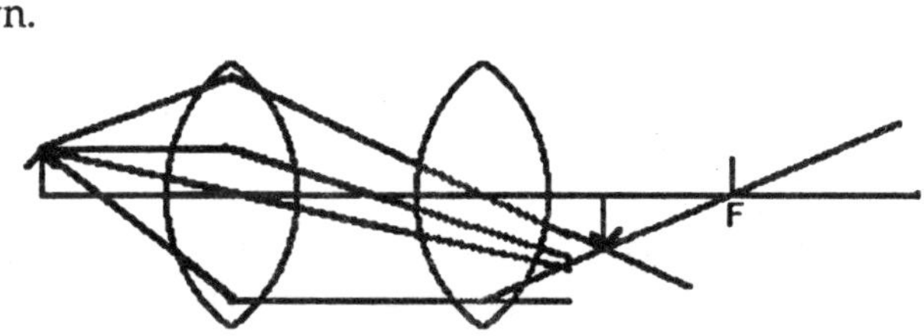

Note that the combination of the two lenses produces a real, inverted image that is closer to the first lens than the image that would have been produced by the first lens alone. In other words, the strength of the combination is greater than the strength of either lens.

Although it is usually easier to consider the effect of the first lens and then add the effect of the second lens, this is not always convenient. For example, consider the case of a person who is *myopic* (near-sighted).The lens in a myopic eye focuses a relatively distant object in front of the retina (rather than on it) as shown below. The green rays show how the image of the red arrow is formed by the lens in the eye acting alone. (The image is real, inverted and smaller than the object – the usual situation for this configuration.)

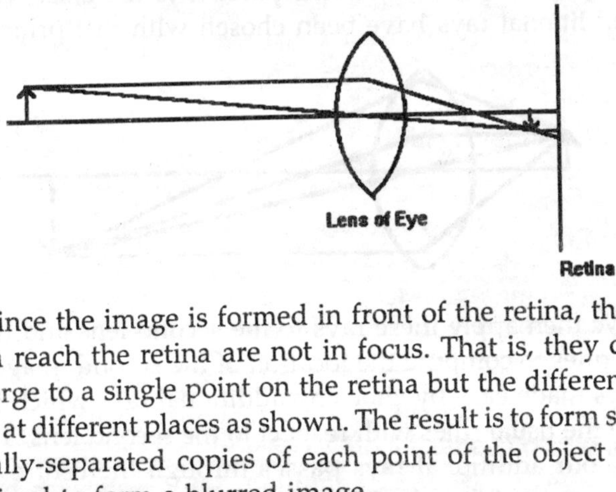

Since the image is formed in front of the retina, the rays which reach the retina are not in focus. That is, they do not converge to a single point on the retina but the different rays strike at different places as shown. The result is to form several spatially-separated copies of each point of the object which combined to form a blurred image.

The solution is to add a negative lens in front of the eye, which moves the red image formed in front of the retina by the lens of the eye alone so that it is formed exactly on the retina as is shown by the purple image in the figure below.

The position of the object and the configuration of the eye are fixed parameters, so that there are only two adjustable parameters – the focal length of the negative lens and where it will be located. If the person chooses conventional eye-glasses (not contact lenses), then the position of the negative lens is fixed by the shape of the person's face, so that the only adjustable parameter is the focal length of the negative lens. This focal length can be determined using a single ray – the purple ray shown below.

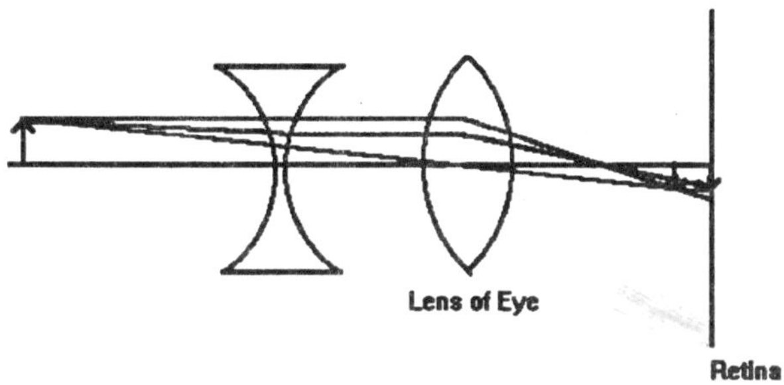

This purple ray leaves the object directed towards the far focal point of the negative lens we are planning to add. It is refracted by the negative lens and emerges parallel to the axis. It then strikes the lens of the eye parallel to the axis and is refracted through its focal point and then strikes the retina as shown. The focal point of the first lens is then determined by simply extending the initial direction of the purple ray and noting where it meets the axis. The distance from this point back to the lens is then the required focal length of the eye-glass lens.

Another way of coming to the same result is to note that the effect of the negative lens is to create a virtual image of the initial object, and to then compute the final image on the retina using this virtual image as an intermediate object. This construction is shown below:

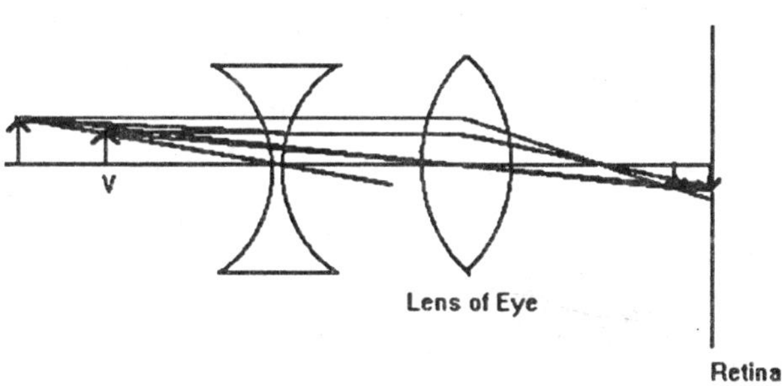

The virtual image produced by the negative lens is shown as "v" in the figure above. Its position is found using two purple rays: a ray headed for the far focal point of the first lens which emerges parallel to the axis (this ray is the same one that was used above) and a ray through the centre of the negative lens which continues unchanged. These two rays appear to diverge from point "v" so that this is the location of the virtual image produced by the negative lens.

Using this virtual image as an object, the image on the retina can be found using the normal rules: a ray emerging from this virtual image parallel to the axis is refracted by the lens in the eye so that it passes through the focal point, and a ray that passes through the centre of the lens in the eye continues in the same direction. These rays are shown in the figure above, and they confirm that the final image is located on the retina as desired.

Another way of describing the effect of the negative lens used in these eye glasses is to note that it works by creating a virtual copy of the object that is closer to the eye than the real object is. The person could accomplish the same thing (without eye glasses) by moving her eyes closer to the real object, and many near-sighted people do just that – they bring relatively distant objects into focus by moving them closer to their face.

Note that this combination of the negative and the positive lens is still positive – the final image is still real and inverted. Only its position has been changed. The size of the final image in this case is about the same with or without the negative lens, but this may not be true if the negative lens is quite strong. People who are very near-sighted see all distant objects as a blur, and usually have to wear glasses all of the time. They get used to the change in the apparent size of objects that is caused by the strong negative lenses; in fact since they can't see clearly without their glasses on, they have no basis for comparing the difference in the apparent size of an object with and without the negative lenses.

These conclusions about the type and size of the final image are not universally true. The image produced by a pair of lenses can be real or virtual, erect or inverted, and need not be the same as the image that would be produced by either lens alone.

There are a number of special cases, where it is easy to compute the effect of a compound lens without all of this work.

1. If the two lenses are touching each other then the effect of the pair is equivalent to the effect that would be produced by a single lens whose power was the algebraic sum of the two powers. (Recall that the power of a lens is measured in diopters and is the reciprocal of the focal length in meters.) Thus in our first example above, the power of the two lens combination is the sum of two positive quantities, which is a numerically larger value. The equivalent focal length of the combination is smaller than the focal length of either lens by itself, so that the resulting image is closer to the first lens as can be seen from the figures. Likewise, the algebraic sum of the strengths of the two lenses in the second example is smaller than the strength of the lens in the eye alone (since the lens we have added has a negative focal length and therefore a negative strength). The result is weaker lens – one whose focal length is greater than the focal length of the lens in the eye alone. The image therefore moves further away from the lens, as is required in this case to put the image on the retina.
2. If the two lenses are separated by a distance d which is not too much greater than the sum of the two focal lengths, then this combination behaves as a single lens whose power is approximately equal to

$$P = \frac{1}{f} = \frac{1}{f_1} - \frac{1}{d - f_2}$$

where f_1 and f_2 are the focal lengths of the two lenses by themselves. If d is much smaller than the focal length of the second lens, then it can be ignored, and this expression reduces to the simple sum of strengths relationship described in the previous paragraph. Note that the equivalent focal length of two positive lenses need not be positive.

The effect of wearing eye-glasses can be estimated using the first rule for contact lenses; although the first rule is approximately correct even for normal (non-contact) configurations, the second rule usually gives a more accurate result for this configuration.

Chapter 8

Aberrations

The graphical methods that we have used to construct images are based on approximations that are never exactly true. Images formed by real lenses and mirrors therefore are never exactly identical to the predictions of the simple paraxial ray methods that we have developed. (This discrepancy is due to the limitations of the paraxial ray approximation, and is present even when the algebraic form of the ray tracing rules is used.) The differences between the prediction of the paraxial ray method and the actual image are called aberrations. These aberrations are usually divided into 5 broad classes:

- *Spherical aberration*: Although our ray-tracing model assumed that all rays parallel to the axis are reflected or refracted so that they pass through a single focal point that was a characteristic of the lens or mirror, this is actually not the case. This assumption gets worse as the distance of the ray from the axis increases. The difference between the focal points for rays that are close to the axis and for rays that strike the lens near its edge is called spherical aberration. Positive spherical aberration means that rays near the edge of the lens have an effective focal point that is closer to the lens than rays that strike the lens near the axis. Negative spherical aberration means that rays near the edge of the lens have an effective focal point that is at a greater distance from the lens than rays that strike the lens near the axis.

Since the effective focal point determines the position of the image for any object, the fact there are several different focal points for the different kinds of rays means that there are several "ghost" images formed by the lens – the "real" one determined by the location of the paraxial-ray focal point, and a number of other images that are formed at different distances depending on where the rays that form them struck the lens. A screen placed at the point where we expect the paraxial ray image therefore also shows these other images not quite in focus, since they are really being formed elsewhere.

Spherical aberration obviously increases with the diameter of the lens, and it can be minimized by limiting the opening of the lens so that only rays in the paraxial region can pass through it.

- *Coma*: This aberration affects rays that come from an object (or a part of an object) that is not at the centre of the lens. The aberration is caused by the fact that the magnification of a lens (that is the size of the image for a given object) is different for rays that strikes the lens in the paraxial region and for rays that strike the lens near its edge. When coma is positive, the image of an object produced by off-axis rays is slightly larger than the image produced by paraxial rays; when coma is negative, the image produced by the off-axis rays is slightly smaller.

Unlike spherical aberration which affects where in space the different images are formed, coma results in several different images of different sizes all of which are in focus on the same screen. The result is a bright image formed by the paraxial rays and a series of smaller (or larger) images formed by the rays that hit the lens far away from the axis. The result often looks a bit like a comet – a clear image with a fuzzy tail oriented along the screen and composed of weaker images formed by the off-axis rays.

There is a defect in the eye which produces an image that looks like coma, but is actually caused by a different problem.

- Astigmatism. This is also an off-axis effect. When an object point is quite far off of the central axis of the lens, the effective radius of curvature (and hence the effective focal length of the lens) in one direction is not the same as in a perpendicular direction. Think of circles of constant longitude on the Earth, which always have the same circumference at any latitude and circles of constant latitude, whose circumference decreases as you move away from the equator. Again, the result is to produce a slight difference in the focal length for rays that are in the longitudinal plan of the lens and for rays that leave this plane. Thus a point on an off-axis object has two image points – one for the rays that strike the lens in the plane of the object and the central axis of the lens and another for rays that strike the lens perpendicular to this plane. At either of these two image points, the rays forming the other image are somewhat out of focus, and often form a small line segment. At intermediate points between the two image points the rays combine to form an image that sometimes looks like a small + sign.

There is also a defect in the eye called astigmatism. It produces a similar effect but is caused by a related, but different problem.

- *Curvature of Field*: Even if all of the effects discussed above could be eliminated, this effect would remain. It arises because the image plane is not really a plane but a spherical surface whose centre is located at the same distance from the lens as the image plane. In the paraxial approximation, the rays passing through the lens strike this image surface near its centre where the curvature of this surface is not great, and the approximation that it is a flat plane is not too bad. However, this approximation gets worse as the rays deviate from the centre of the lens. The effect of this field curvature is to produce a blurring the edges of the image.

It is possible to correct for this effect using a combination of a positive and a negative lens that are very close together, and this is usually done in camera lenses.

- *Distortion*: This effect is caused by a variation in the magnification of the image across the field of view. That is, some parts of the image are magnified more than others, so that the overall effect is to produce a distorted version of the object, which is still in focus everywhere. This aberration does not affect the focus of the image, but it does affect the accuracy with which the image represents an accurate representation of the object. Distortion is very noticeable at the edges of a convex mirror, for example, where the magnification is very different from the magnification in the paraxial region.
- *Chromatic aberration*: This is caused by a variation in the index of refraction of a lens with the wavelength of the incident light. It is in addition to all of the other aberrations discussed above, which are caused by the inadequacy of the paraxial ray approximation and have nothing to do with the colour of the incident light. It is only present in lenses – mirrors obviously have no chromatic aberration because they operate by reflection rather than refraction.

The most noticeable effect of chromatic aberration is that images of objects illuminated by white light show coloured bands around the edges because the different colours in the white light focus at different points in the image plane. This effect is in addition to all of the other aberrations discussed above.

It is possible to correct for chromatic aberration by using two lenses made of different materials so that variation in the index of refraction of one lens is cancelled by the opposite variation of the other one. Such doublets are called "achromats" and are commonly used in camera lenses.

CHROMATIC ABERRATION

We have seen that both mirrors and lenses suffer from spherical aberration, an effect which limits the clarity and sharpness of the images formed by such devices. However, lenses also suffer from another type of abberation called *chromatic abberation*. This occurs because the index of refraction of the glass in a lens is different for different wavelengths. We have seen that a prism refracts violet light more than red light. The same is true of lenses. As a result, a simple lens focuses violet light closer to the lens than it focuses red light. Hence, white light produces a slightly blurred image of an object, with coloured edges.

For many years, chromatic abberation was a sufficiently serious problem for lenses that scientists tried to find ways of reducing the number of lenses in scientific instruments, or even eliminating them all together. For instance, Isaac Newton developed a type of telescope, now called the Newtonian telescope, which uses a mirror instead of a lens to collect light. However, in 1758, John Dollond, an English optician, discovered a way to eliminate chromatic abberation. He combined two lenses, one converging, the other diverging, to make an *achromatic doublet*. The two lenses in an achromatic doublet are made of different type of glass with indices of refraction chosen such that the combination brings any two chosen colours to the same sharp focus.

Modern scientific instruments use *compound lenses* (*i.e.*, combinations of simple lenses) to simultaneously eliminate both chromatic and spherical aberration.

Chapter 9

Camera Lenses, Focusing on the Object

The lens of a camera must form a real image of the object that is being photographed – the rays of light from a point on the object must actually strike the film at a single point in order to trigger the chemical reaction that records the image. Although the lens of a camera is a complex multi-element design to minimize aberrations, its general properties are equivalent to a positive lens of comparable focal length. We can therefore understand the performance of a camera lens.

The ray-tracing rules we have established define a unique relationship between the focal length of the lens and the positions of the object and the image. In other words, when a lens of some focal length is placed at a given distance from the film plane then this distance and the focal length of the lens determine the distance to an object that is in sharp focus. In principle, objects at any other distance are not clearly focused. In order to bring an object at some other distance into clear focus, it is necessary either to move the lens or to change its focal length. Either of these changes is possible, and both are supported by more expensive cameras. Simpler and cheaper cameras usually have the ability to move the lens but not to change its focal length, while neither the focal length nor the lens position can be changed on the very simplest cameras.

In order to get a feeling for how much the lens must be moved to bring different objects into focus, the following table is computed using the algebraic form of the ray-tracing equations for a single positive lens.

Distance of the object in front of the lens in units of the focal length	Position of the image behind the lens in units of the focal length
10	1.11
100	1.01
1,000	1.001
10,000	1.0001

For example, suppose we are using a lens whose focal length is 50 mm, which is a typical value for the lens on a 35 mm camera. The first entry in the table above says that an object 500 mm (just under 20 inches) in front of the lens (10 × 50) is imaged 55.5 mm behind the lens (1.11 × 50). If this object is to be imaged on the film plane, the lens must be located 55.5 mm in front of the film. The next entry says that an object that is 5000 mm (5 m) in front of the lens (about 16 feet) will be imaged 50.5 mm behind the lens. The lens must therefore be moved closer to the film plane in order to focus on this more distant object. The same thing is true for the other entries – the lens must be moved closer to the film plane as the distance to the object increases. Note that the changes become smaller and smaller as the distance to the object increases. In order to bring objects from 20 inches to infinity into clear focus, this particular lens must be able to move about 5.5 mm – from 55.5 mm in front of the film to essentially 50 mm in front of the film plane.

The lens motion required to bring these objects into focus is surprisingly small, and it is not difficult to design a camera that allows the lenses to move by these amounts relative to the film plane. On the other hand, the position of the image changes more and more rapidly as the objects gets closer to the lens, and most lenses cannot move far enough to focus on close-in objects for that reason. For example, an object that is 5 focal lengths in front of the lens has an image that is 1.25 focal lengths behind the lens, while an object that is 2 focal lengths in front of the lens has an image that is 2 focal lengths behind it. These motions cannot be accommodated by standard lens mounts, although there are special-purpose mounts that can do this.

DEPTH OF FIELD

We showed how the position of the lens had to be moved relative to the film on the image plane in order to bring objects at different distances in front of the lens into sharp focus. However, simply changing the position of the lens is not adequate for real-life situations, since anything more complicated than the very simplest of objects will have components that are at different distances from the lens. In principle, all of these different components cannot be brought into sharp focus on the film plane simultaneously. The following figure illustrates the problem.

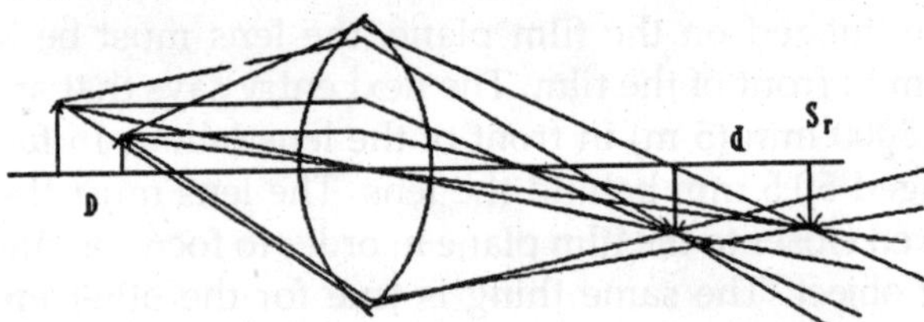

If the lens is adjusted so that the image of the red object is in sharp focus (that is, the lens is adjusted so that the film plane is at distance S_r behind the lens) the rays from the blue object strike the image plane in an extended region rather than at a single point. Depending on the goal, the lens can be moved to bring either of the two objects into sharp focus (but not both simultaneously) or it can be adjusted to a point midway between the two images so that both objects are about equally blurred.

The magnitude of this problem is determined by two independent factors. The first is the angle at which rays from the lens converge to the point of sharpest focus. This angle determines how much blurring will result from a given difference in the positions of the images from the different parts of the object. (The rays diverge on the far side of the point of sharpest focus at the same angle, so that the blurring is equally large at some point in front of the actual focus or at a point an equal distance behind it.) This angle is illustrated in the following figure.

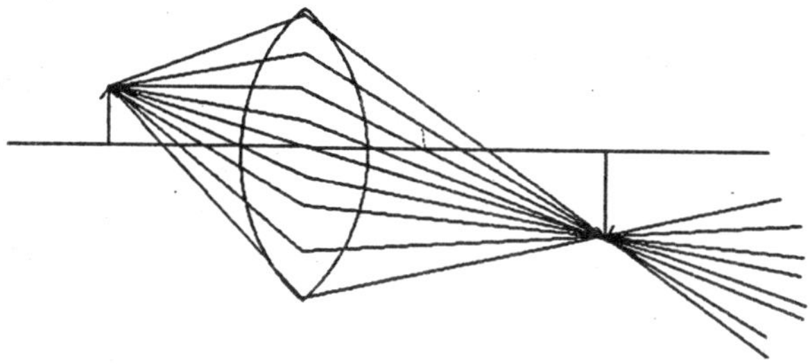

The size of this angle is determined by the ratio of the diameter of the lens to the distance between the lens and the image. Since objects are usually many focal lengths in front of the lens, the distance between the lens and the image is approximately equal to the focal length of the lens. The ratio of the diameter of the lens to its focal length has appeared before as the parameter that governs the applicability of the paraxial ray approximation, and its inverse is the f-number.

The magnitude of this angle is clearly determined by the extreme rays – the rays that strike the lens at its edges, so that the problem can be reduced either by increasing the focal length of the lens (leaving everything else unchanged) or by inserting a stop or diaphragm that blocks the edges of the lens and prevents these extreme rays from reaching the image. Since the f-number is the ratio of the focal length to the diameter, reducing the effective diameter of the lens leaving the focal length unchanged increases the effective f-number of the lens. (Increasing the focal length leaving the diameter unchanged obviously has the same effect.) This increase in f-number reduces the angle at which the rays converge to (and diverge from) the point of sharpest focus and reduces the blurring for any given configuration.

The second factor that governs the magnitude of this blurring is the relationship between D, the distance between the two points in the object space, and d, the distance between the corresponding images. Decreasing d, keeping everything else the same, will decrease the size of the problem because the same angular divergence of the various rays produce a smaller divergence of the rays on the image plane.

Therefore: the blurring due to depth of field can be reduced by using a larger f-number and leaving everything else unchanged.

For a given distance D between two points on the object, the corresponding distance between the two images decreases as the objects move further away from the lens. This point is illustrated in the following table, which gives the variation in the ratio d/D as the object moves further away from the lens.

Distance of the object from the lens in units of the focal lengt	d/D
5	0.06
10	0.01
100	0.0001
1000	10^{-6}

For example, if an object is 5 m (about 16 feet) in front of a lens whose focal length is 50 mm (a distance of 100 focal lengths), the ratio of d/D is 0.0001 – the images of two objects are spaced only 0.0001 times the distance between the two objects themselves. As you can see from the table, the fraction d/D decreases as the *square* of the distance from the object to the lens.

Therefore: the blurring due to depth of field will be reduced as the object moves further away from the camera, leaving everything else unchanged.

The final factor in estimating the depth of field of a real lens is the fact that no lens can focus the rays from a single point on an object to a single image point. Even the best lens produces a somewhat blurry image because of residual uncorrected aberrations, because of scattering of light in the lens, and because of dispersion (the fact that the ratio of the wavelength of the light to the size of the aperture is not exactly zero so that light always has some residual wave-like character).

The practical depth of field of a lens is then set by the balance between the imperfections in the size of the image point due to all of these problems and the additional blurring caused by depth of field. That is, the practical depth of field of a lens is set as the point at which the blurring due to the depth of the real object is not greater than the blurring that is inherent in the performance of the lens itself when it is imaging a perfectly flat object. For any given lens and object, the two factors above play important roles – the depth of field can be increased by increasing the f-number of the lens or by moving the camera further away from the object.

The following table shows the actual depth of field for a typical lens. That is, the table shows the range of object distances over which the image is essentially in "perfect" focus because the blurring due to the differences in the object distances is not greater than the blurring due to the other imperfections in the optical system. This lens has a focal length of 50 mm and a maximum aperture of f/1.7 (In other words, the maximum aperture is 50/1.7=29.4 mm). These parameters are typical of the lens that is used on an ordinary 35 mm camera.

Object distance (m)	**Depth of field in meters at f-number**				
	1.7	4	8	11	16
5	4.5-5.5	4-6.5	3.4-9.6	3-15	2.5-175
10	8.3-12.6	6.8-20	5-500	4-infinity	3.4-infinity
100	47-infinity	20-infinity	10-infinity	7-infinity	5-infinity

Although the details of this table will vary from lens to lens, the general features are universal. In particular, when a lens is used at a small aperture opening corresponding to f/11 or f/16 and is focused on an object 10 m (about 30 feet) away, almost all objects (except for those that are quite close to the camera) will be in focus. This is how most cheap cameras are configured – they operate at a fixed lens aperture of f/11 or f/16 with a fixed focus lens adjusted so that objects about 5-10 m away are in sharpest focus. The depth of field at this small aperture is so large that almost everything else is acceptably in focus without the need for any adjustments.

Camera Lenses, The Effect of Focal Length

The focal length of the lens used in a camera affects two aspects of the image: the field of view and the magnification.The field of view is the size of the object that can be contained on the film and the magnification is the ratio of the size of the image to the size of the object that produced it. These two quantities are related to each other – increasing the magnification while leaving the size of the image unchanged inevitably decreases the field of view (since the increase in the magnification implies that the same sized image will be produced by a smaller object), and vice-versa.

The field of view is usually expressed as an angle. There are a number of different ways of estimating this quantity, but all of them yield about the same value. Since we are going to use the approximation that the camera lens is a simple single thin lens, the calculation will not be exactly correct anyway. As with previous calculations, the agreement between this simple model and a real-lens is surprisingly good.

If the object is not too close to the lens, then the image is always formed approximately at the focal distance back from the lens. The size of the image on the image plane is usually limited by the size of the film, and the angular size of the field of view is given by those rays that enter the lens at its centre (and so are not deviated) and which then strike the edges of the film as shown in the following figure:

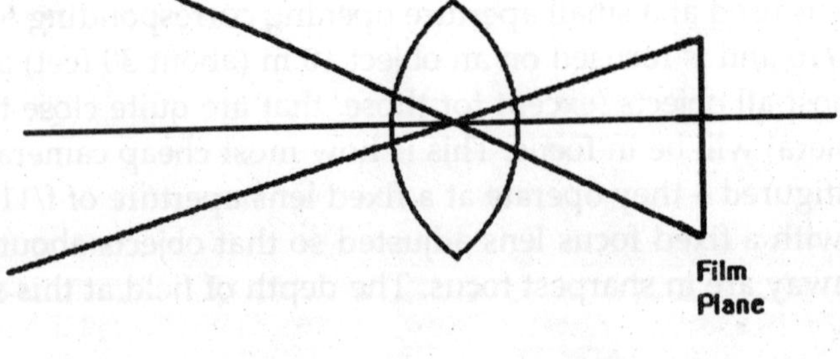

The angles on the object and image side of the lens are the same, so that the tangent of one-half of the angle of the field of view is approximately the ratio of the half-width of the film to the focal length of the lens. For example, the half-width of 35 mm film is 18 mm. The horizontal field of view of a 50 mm lens is then twice the arc-tangent of 18/50 or about 40 degrees. The film is not square, so that the field of view in the vertical direction is not the same. The half-height of the film is only 12 mm, so that the vertical field of view is twice the arc tangent of 12/50 or about 27 degrees.

(Another method for calculating the field of view would be to consider those rays that pass through the first focal point on the object side of the lens. These rays are refracted so that they emerge parallel to the axis on the image side of the lens. The angular spread of these rays provides an alternate method for estimating the field of view. This estimate is essentially the same as the estimate using the method described above unless the object is quite close to the lens. The two estimates differ in this situation because the lens must be moved away from the film plane in this situation so that the distance from the lens to the film in the first figure is somewhat greater than the focal length.) See the following figure, which compares the field of view calculated in this way with the calculation above when the object is far away.

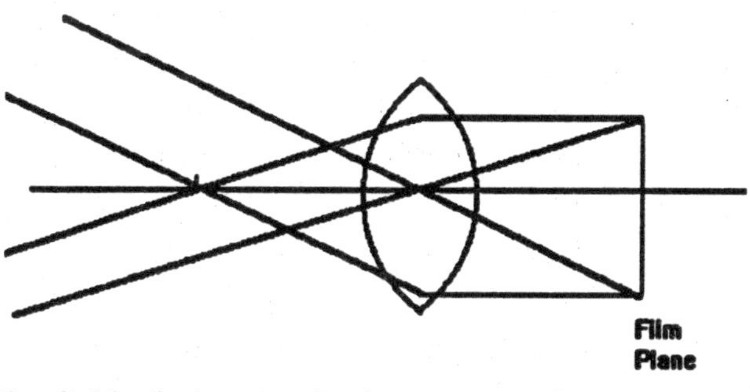

The field of view angle decreases with increasing focal length, but the relationship is through the arc tangent function and is not exactly linear (except at very small angles). The horizontal field of view of a telephoto lens whose focal length was 135 mm would be twice the arc tangent of 18/135 or about 15 degrees.

The magnification of a lens can be estimated using the same sort of analysis. The following figure shows the image produced by an object.

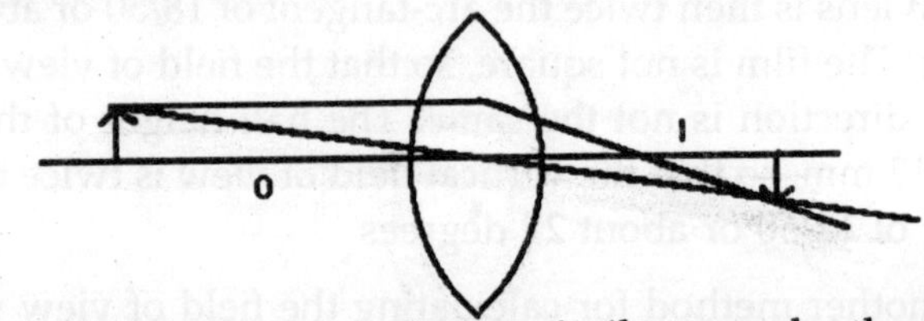

The two shaded triangles are similar to each other, so that the corresponding sides are proportional. The magnification is the ratio of the size of the image to the size of the object that produced it. Since the triangles are similar, that ratio is also equal to the ratio of the distance between the lens and the image ("I" in the figure) and the distance between the lens and the object ("O" in the figure). Since the distance "I" is approximately the focal distance, the magnification of a lens for a fixed object depends directly on the focal length and is equal to the ratio of the focal length to the distance of the object from the lens.

The magnification increases directly with the focal length when the object, the position of the camera, etc. are unchanged. Thus the same object appears larger when imaged by a camera with a telephoto lens because the focal length of that lens is larger than the focal length of a normal lens.

Although a film image is purely two-dimensional, the variation in the magnification of a lens with object distance can distort perspective – that is, how we estimate the size of something in an image by comparing it to other things in the same image. For example, if an object is sufficiently large so that the distances between different parts of the object and the lens are significantly different, then the parts that are closer to the lens will have a greater magnification than the parts that are further away. This effect can be seen when photographing a tall building from a point near its base. The bottom of the building is significantly closer to the camera than the top, is magnified by a larger amount and appears disproportionately too large as a result.

FOCUSING THE LENS, RANGEFINDERS

In cameras where the lens can be moved to bring the object into focus on the film plane, there is usually a mechanism that indicates when the proper position of the lens has been reached. One system that is used for providing this indication is the optical rangefinder. A number of different versions of this device have been developed, but most of them are based on the design shown in the following figure. (The optical components that are shown in the figure are usually completely independent from the lens that is used to take the picture.)

The goal is to focus the camera on the object by observing its image on the viewfinder screen.

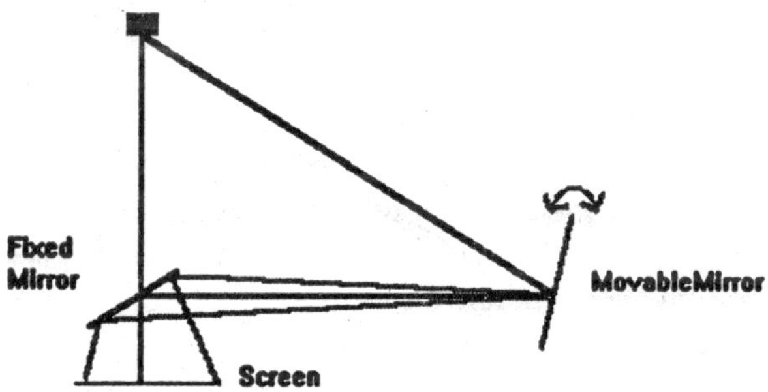

The user sees two images of the object on the screen inside of the camera. One image is produced by rays that travel along the direct path from the object to the screen as shown in purple. Rays traveling along this path strike the screen after passing through the fixed mirror, which transmits 50per cent of the light that is incident on it and reflects the other 50per cent. The second image is produced by rays traveling along the blue path. These rays are reflected twice before hitting the screen: once by a movable mirror that can rotate about an axis perpendicular to the figure and the second time by the fixed mirror. As the movable mirror is rotated, the blue image moves across the screen, and there is one position of the movable mirror for which the blue image exactly coincides with the purple image. (At other positions, the rays are reflected to other positions on the screen as shown by the orange rays, and the user sees a displaced image.) The position at which the two images overlap is determined by the geometry of the mirrors and by the distance of the object from the camera, so that the angle of the movable mirror can be calibrated to read the distance to the object directly.

The lever that rotates the movable mirror is coupled to the system that positions the lens, so that rotating the mirror until the two images overlap simultaneously adjusts the lens to focus on the object at that distance. This "coupled rangefinder" system is widely used on moderately-priced cameras.

The rangefinder is most sensitive to objects that are relatively near to the camera, and becomes less sensitive as the object moves further away because the corresponding changes in the angular position of the mirror become smaller and smaller at greater distances. This is usually not a serious problem, because the depth of field of the lens is also increasing as the object moves further away, so that the focusing becomes less critical at these distances.

Focusing the Lens,Bi-Prisms

A second method for detecting when the lens has been focused on the desired object uses a pair of prisms mounted in opposite directions on the image screen. This configuration is sometimes called a biprism, although that name is also used for other configurations of two prisms that have nothing to do with focusing a camera.

The method depends on the properties of a single thin prism as shown in the following figure.

A user whose eye is looking down through the prism from above sees an object below the prism as if it were located along the straight path shown by the blue line. In fact, the ray that reaches the user's eye was refracted twice by the prism and actually followed the path shown by the green line. The angle of divergence between the blue path and the green path is a constant value – it depends only on the index of refraction of the prism and on its shape.

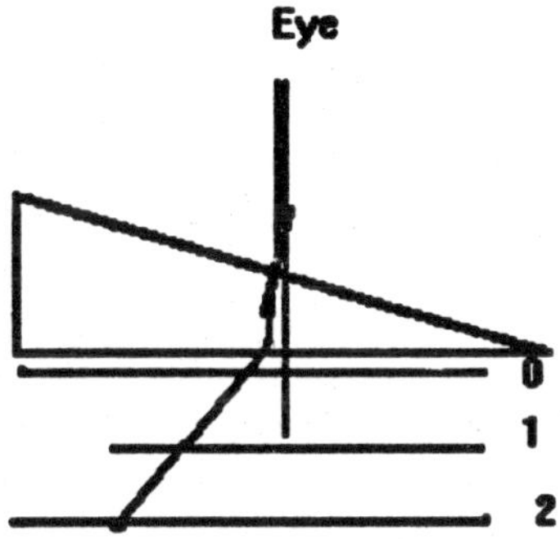

If the user is looking at an object located at position "0" then the user actually sees a point on the object that is to the left of the straight line path because the ray that actually enters the user's eye was refracted as shown. As the object moves away from the bottom of the prism towards positions "1" and "2" the point on the object whose rays strike the user's eye moves further and further to the left as shown in the figure. The user thinks that the object is sliding to the right, since parts of it that were further and further to the left are coming into the view point which looks to be directly under the user's eye. On the other hand, when the object moves upward so that it is directly underneath the prism, the difference between the green ray and the blue ray is very small so that the user sees an object point that really is directly on the straight line of sight. (The difference between the blue path and the green path has been exaggerated in this figure by separating the two rays for clarity.)

We now combine two identical prisms of this type with one of them slanted to the right and one of them slanted to the left. We arrange the configuration so that the top half of the object sends it rays through one of the prisms and the bottom half of the object sends its rays through the other one.

As the object moves away from the bottom plane of the prisms, the two halves of the image appear to move in opposite directions, since the prisms tilts are opposite. Conversely, as the object moves closer to the bottom plane of the prism, the two halves of the image move closer together. If the prisms are thin and no too tilted, the two halves of the image will line up exactly with each other when the object plane is exactly at the common base plane of the two prisms.

In practice, the two prisms are located on the imaging screen (this screen is where the main camera lens forms its image while the picture is being composed), and the image on this screen serves as the object for the two prisms. When the image produced by the main camera lens is located exactly on the imaging screen, the two halves of this image transmitted through the two prisms appear lined up. When the image produced by the main camera lens is not focused on the imaging screen, the two halves of the image transmitted through the two prisms are displaced in opposite directions, and the image seen by the user appears to be cut in half at its centre with the upper and lower halves displaced relative to each other.

Some imaging screens contain additional very small pairs of these prisms called micro-prisms. Each pair functions in the same way – the image appears to be smooth and continuous when all of the little pieces are not displaced relative to each other and appears to be broken up into tiny disjoint parts when the image is out of focus.

SHUTTERS

The shutter of a camera has only one function – to control the amount of the time that the film is exposed to the scene. This is necessary for two reasons:

1. If the object is moving or the scene is changing in some other way, then the shutter can be used to limit the exposure to a time interval that is short enough to "freeze" the motion, and
2. The exposure time controls the amount of light that strikes the film to be sure that the film is properly

exposed.Early cameras used leaf shutters that were located in front of the lens or between two elements of a multi-element lens. As the shutter opened, a circular hole was formed that increased in diameter until the shutter was fully opened and then decreased in diameter until the shutter was fully closed again. Since the shutter was not in the image plane, the opening and closing of the shutter affected the entire image in the same way. That is, the opening of the shutter gradually allowed more and more off-axis rays to reach the film plane, but all of the film plane was exposed to the incident light at all shutter positions. The shutter function was often combined with the lens opening adjustment into a single device whose opening speed set the exposure time and whose maximum diameter set the f/number. Note that the shutter has to move faster and faster at shorter exposure times, and this requirement set a mechanical limit on the shortest exposure times that could be obtained.

Although shutters of this type have some theoretical advantages, they are difficult to use in cameras with removable lenses. Since the shutter is an integral part of the lens, both the shutter and its trigger mechanism must be built into all of the different lenses that will be used. They are also difficult to use in single lens reflex cameras, since those cameras use a through-the-lens viewfinder to compose the image. The viewfinder obviously couldn't work if the shutter was closed. All single lens reflex cameras and most other cameras with removable lenses therefore use a focal-plane shutter.

A focal plane shutter is located just in front of the film. It consists of two pieces of opaque material called the "curtain." When the shutter is to open, the first piece of material slides horizontally across the film plane, gradually exposing the film from one side to the other (or sometimes from top to bottom). After a preset time, the second piece of material starts to move across the film plane gradually closing the opening again. At relatively slow shutter speeds (typically about 1/60 s and longer), the second piece of the curtain does not start moving until after the first piece has reached the other end of the film. The entire image area is therefore exposed for some period of time before the second curtain starts to close the opening. At higher shutter speeds, the second curtain starts moving to close the opening before the first curtain has finished opening it. The result is a moving "slot" which exposes different parts of the film at slightly different times. In most cameras, the curtains move at the same speed for all exposure times, and the width of the slot is varied to set the required exposure time. The shortest exposure time is then set by the speed of the curtains and the minimum width of the slot. Most cameras of this type can support exposure time down to about 1/1000 s.

Focal plane shutters cannot be used at all speeds with flash bulbs or electronic flash units. This is because the duration of the flash is very short – typically about 0.001 s. If the shutter is operating in its "moving slot" mode, then only a small piece of the film will receive the light of the flash unit – the piece of film that is being exposed when the flash is fired. In order for the entire image to receive the light from the flash unit, the shutter must be set to a slow enough speed so that it is running in its fully open mode. That is, the mode in which the second curtain does not start closing the opening before the first curtain has finished opening it. The flash fires in this instant after the first curtain is fully open and before the second one starts to close. The maximum speed of the shutter in which this is true varies from one design to another but is typically about 1/60 s, and this speed is usually indicated on the shutter speed selector in some way.

FILM SPEEDS AND EXPOSURE CALCULATIONS

The chemical response of the film is determined by two parameters: the amount of light that strikes the film and the sensitivity of the chemicals to that exposure. The intensity depends on three parameters:

1. The diameter of the lens, which determines how much light enters the camera,

2. The focal length of the lens, which determines the size of the image that is produced by the object and therefore how the energy captured by the lens is distributed over the film plane,
3. The amount of time the shutter is open

The amount of light that is captured by the lens depends directly on the area of the lens opening, which varies as the *square* of the diameter. Likewise the energy striking a given area of the film varies *inversely* as the *square* of the focal length, since the focal length affects both the length and the width of the image, and it is the product of these two quantities that determines the area. Increasing the focal length increases the length and width of the image linearly, and increases the area (and therefore decreases the intensity on the film) by the same quadratic factor. The shutter time enters linearly – doubling the exposure time doubles the intensity of the light striking the film plan.

Since the diameter of the lens and its focal length have the same dependence on the amount of light striking the film, and since the ratio of these two quantities (f/D) is the f/number, it is possible to combine these two effects into the statement that the amount of light striking the film varies as the square of the f/number, with increasing f/numbers implying corresponding decreases in the amount of light.

There are 3 methods for specifying the "speed" of a film – that is, the relationship between the amount of light that strikes the film and the density of the chemical reaction (the "exposure") that results. All of these speed specifications are relative, in that they specify the sensitivity of a particular film compared to the sensitivity of some standard. These 3 methods are the ASA (American National Standards Institute) scale, the DIN (Deutsche Industrie Norm) scale and the ISO (International Standards Organization) scale. All of these scales are based on the following figure, which shows the chemical response of the film as a function of the intensity of the light that strikes it.

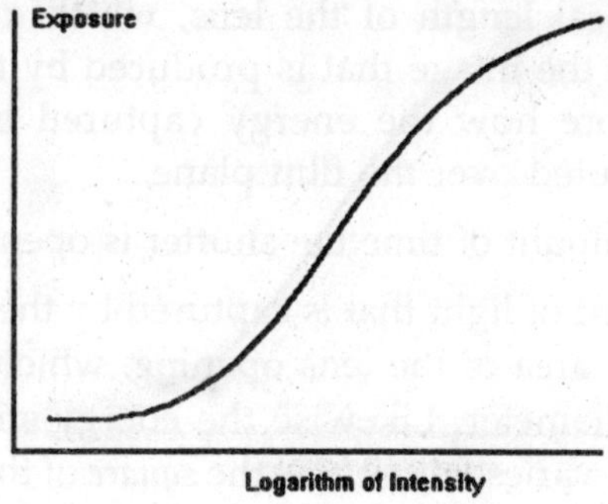

Note that the scale on the x-axis is logarithmic rather than linear. Each division on the x-axis therefore represents a certain multiplicative increment rather than an additive increment as would be true for a linear scale. The most common multiplicative increment is an increase (or decrease) in the intensity by a multiplicative factor of 2, and this same multiplicative factor is used in calculating exposures as discussed below.

Using this curve for any film, the testing laboratories can assign a relative speed number to it. The ASA index is linear: a film rated as ASA 50 is twice as sensitive as one rated as ASA 25, and requires half the light for the same chemical exposure. The DIN scale is explicitly logarithmic. An increase of 3 in the DIN rating corresponds to a factor of 2 increase in sensitivity. Thus DIN 21 is twice as sensitive as DIN 18. The ISO index is also logarithmic, but is computed using a slightly different method than is used by the DIN system. The results are basically the same as for the ASA method. The correspondence is that ASA 100 is equivalent to DIN 21, which is equivalent to ISO 100/21 degrees. Using these definitions, ASA 200 would be DIN 24, ASA 400 would be DIN 27, etc. Most film is characterized using both the ASA and DIN values.

Although the details vary from one type of film to another, there are three general characteristics that all films share:

1. A relatively flat region at very low light intensities. Nothing much happens in this region over a significant range of intensities. This region is

responsible for what is often called "reciprocity failure," which means that beyond some point, decreasing the intensity of light that strikes the film results in almost no change in the exposure of the chemicals. In other words, a wide range of exposure times produces about the same (underexposed) image.

2. The central region, which is roughly a straight line on the logarithmic display. Equal changes in intensity produce roughly equal changes in the density of the film exposure.
3. A relatively flat region at very high light intensities. The film is saturated at these intensities and further increases in intensity cannot produce additional exposure. The picture will be "over-exposed" with too little contrast, because even the darkest area of the image produced almost the same chemical reaction as the brightest area.

Most photography is done in region (2), although the other regions are used for special effects. For example, region (1) is often used for long exposures at night. A film that is intended for photographing printed text is designed to have a very narrow region (2), so that the response of the film is either in region (1) or in region (3). That is, it is either fully exposed or not exposed at all – consistent with the nature of the material, which is black and white with no intermediate shades of gray.

Digital cameras have the same sort of response curve, but it is usually not published. The advertising for digital cameras tends to concentrate on the number of pixels in the image instead. Although this is an important parameter since it sets the resolution of the camera, the sensitivity of the digital recording medium, expressed in units comparable to an ASA number would be very useful in determining what level of light would be required to take a picture. Many of the earlier digital cameras had sensitivities comparable to film rated at ASA 100 – not very sensitive by film standards.

Since the exposure in region (2) varies smoothly with intensity, it is possible to trade-off the various parameters that affect the intensity while preserving more or less the same exposure. For example, decreasing the speed of the shutter by a factor of 2 can be compensated for by decreasing the f/number by 1.4 (the square root of 2) thereby increasing the amount of light that strikes the film by the same factor of 2. Thus there are pairs of f/numbers and shutter speeds that give essentially the same exposures. The choice among these equivalent settings is driven by other considerations. For example, using a larger f/number will increase the depth of field but require a longer exposure time, whereas using a faster shutter speed will capture a static image of an object that is moving rapidly but will require a smaller f/number to capture adequate light during this shorter exposure time.

Although any number of values could be chosen, it is common to use values both for the shutter speed and for the f/number that change by a factor of 2 starting at f/1 and a shutter speed of 1 sec. Thus the common choices are:

F/number values: 0.7, 1, 1.4, 2, 2.8, 4, 5.6, 8, 11, 16, 22, ...

Shutter speeds: 1, 1/2, 1/4, 1/8, 1/15, 1/30, 1/60, 1/125, 1/250, 1/500 ...

In both series, the values are rounded to convenient fractions.

Within the linear region of the exposure curve, 1/60 at f/8 results in the same exposure on the film as 1/30 at f/11, etc. That is, an increase of a factor of 2 in the time the shutter is open (from 1/60 to 1/30) is compensated by decreasing the size of the lens opening by the square root of 2 from f/8 to f/11. Furthermore, changing the ASA rating of the film by a factor of 2 would require a corresponding compensation either by a change in shutter speed of a factor of 2 or a change in f/number by a factor of 1.4.

Since there is usually more than one acceptable combination of f/number and exposure time, automatic cameras usually have several choices: an "aperture priority" mode in which the user can choose the f/number and the camera then chooses the correct exposure time and a "shutter priority" mode in which the user chooses the exposure time and the camera chooses the f/number to match it. The camera tries to keep both parameters in the middle of their respective ranges by default.

When a camera is used with an electronic flash, then most of the light usually comes from the flash, and the duration of the flash (which is usually very short) sets the effective exposure time. The shutter speed is pretty much irrelevant in this situation, provided that the motion of the shutter and the firing of the flash are synchronized.

THE OPTICS OF THE EYE

The Lens

Although the lenses of the eye are not simple spherical lenses, they are thicker in the centre than at the edges. Therefore, they are positive and converging, and form a real, inverted image on the retina – the optically sensitive screen that is located at the back of the eye. The image on the retina is formed by two lenses – the cornea, which is a converging lens whose focal length is fixed, and the eye-lens, which is a positive lens whose focal length can be varied by changing its shape. The focal length of the combination varies among different people, but is typically about 20 mm.

The optical strength of the cornea is much greater than the strength of the eye-lens, and most of the focusing of the lens combination in humans is due to this part of the optical system. The focal length of the cornea depends on its internal index of refraction and on the fact that the material in front of it is air. The focal length is therefore very different when the cornea is in water, since the refraction at the front surface of the cornea is much less in this situation. (This is why it is easier to see underwater when you are wearing goggles, since the goggles produce an air space in front of the cornea.) Fish and other animals that see in the water therefore have eyes whose lenses are much more curved and bulging than human eyes.

In all cases, the index of refraction of the lens material is much closer to the index of refraction of water, so that a lens of the same shape has a much smaller optical strength. Lenses that must operate in water are therefore much more curved than a lens of equivalent focal length designed to work in air.

Neither surface of the eye-lens is exposed to air, and its focal length is determined by its index of refraction and by the indices of refraction of the materials that surround it. The index of refraction of the eye-lens is not homogeneous but is smallest at the surface and increases somewhat with depth.

The eye focuses on objects at different distances by changing the focal length of the eye lens. This change is realised by changing its shape – decreasing its radius of curvature (and therefore shortening its focal length) as the object comes closer to the eye. The range over which this works varies among different people, but is typically from infinity to the *near point*, which is about 25 cm (10 inches) for most people. This variation in the focal length of the eye-lens is called *accommodation*, and this ability to vary the focal length of the eye-lens decreases slowly with age. Since the eye is focused on distant objects when the muscles are relaxed, the decreases in the strength of the muscles and in the flexibility of the lens with age generally means that people lose the ability to focus on nearby objects as they grow older. They therefore often tend to become *far-sighted* as a result.

The Iris and the Pupil

The iris is a circular diaphragm, and its colour is what we call the colour of a person's eyes. The colour of the iris has no role in the optical characteristics of the eye. The central opening of the iris is called the pupil. The diameter of the opening changes in response to changes in the ambient light level. It varies by about a factor of 4 in people: from about 2 mm in diameter in bright light to about 8 mm in diameter in the dark. The variation is much larger among nocturnal animals, such as cats. The f/number of the human eye therefore varies from about f/2.5 (20/8) in dim light to about f/10 (20/2) in bright light. The variation in ambient light level is much larger than this factor of about 16 in light-gathering ability, and the eye uses other mechanisms to adjust to changes in the intensity of the ambient light. However, the increase in f/number as the ambient light increases is important in increasing the depth of field of the optical system, and it is one of the reasons that it is easier to read and do other tasks that require precise focusing when the light is brighter. As we have discussed previously, many other optical aberrations are also smaller at larger f/numbers, since the paraxial-ray approximation is more accurate in this case.

The largest residual aberration is usually chromatic aberration, which is due to the variation in the index of refraction of the eye lenses with wavelength. When the ambient intensity is high, the use of the cones to process the image tends to minimize this problem by selectively ignoring the blue and violet portions of the image, where the chromatic aberration is the largest, and concentrating instead on the yellow and green portions, which are at the centre of the visible spectrum. However, objects illuminated with light containing lots of violet and yellow (or ultra-violet and blue) can appear fuzzy because the images in the two different colours cannot be brought to a sharp focus simultaneously.

The Retina

The retina is the optically-sensitive portion of the eye. It is the screen on which images are formed, and is connected to the brain via the optic nerve. The retina is composed of about 130 million sensors called rods and cones. The rods are more sensitive to low light levels, whereas the cones operate at higher light levels and are also responsible for colour vision.

The cones are concentrated at the fovea, a point near the optical axis of the eye. Since this point is close to the optical axis of the eye, the various lens aberrations are smallest here, and the image quality is therefore best at this point. In addition, the density of cones is very high in this region, and the resolution of the eye is therefore greatest when the image falls on this area.

The rods predominate at other places on the retina, especially at the edges, which have mostly rods and few cones. The rods are most sensitive to low light levels, but cannot distinguish colour. Therefore perception of colour gets poorer as the ambient light level decreases. Since there are more rods on the periphery of your retina than at the centre, peripheral vision tends to be more sensitive at lower levels of ambient light. The image quality is poorer, however, both because of the lower density of rods in these regions and because the aberrations are relatively larger this far from the optical axis.

The sensitivity of both the rods and the cones can be adjusted in response to changes in ambient light intensity, and this is one of the ways the eye adjusts its sensitivity. These changes are much larger than the change in effective f/number produced by changes in the size of the pupil. This effect is called *adaptation*. It is a relatively slow process – the change in the sensitivity of the retina to a sharp change in ambient light intensity typically takes 15 or 20 minutes.

In addition to being more sensitive at lower levels of ambient light, the rods are also more sensitive to blue light than to red light. If an object is illuminated with only red light, the rods will tend to see the scene as "dark" and will increase their sensitivity as a result, while the cones will tend to see the scene as lighter and will become active because of that. This trick is often used in illuminating the instrument panel of a car at night and in movie theatres. In both cases, the goal is to provide the best features of night vision using the rods and day vision using the cones simultaneously.

CORRECTING COMMON VISION DEFECTS

In order to understand common vision defects, it is helpful to understand how the "normal" eye works. The muscles that control the focal length of the eye-lens are fully relaxed when the eye is focused on a very distant object. The muscles contract and shorten the focal length of the eye-lens to bring nearer objects into focus. The two limits of this range are called the far-point, which is the distance of an object that is in focus when the muscles are relaxed, and the near-point, which is the distance at which an object is in focus when the muscles have shortened the focal length of the eye-lens as much as possible.

The distance between the eye-lens and the retina is about 20 mm. When an object is very far away from the eye, the image is located essentially at the focal point. Therefore the focal length of the cornea and the eye-lens should be about 20 mm when the muscles of the eye are relaxed. Since the strength of a lens is the reciprocal of its focal length in meters, the strength of the cornea and the eye-lens in this situation is about 1/0.020 = 50 diopters.

When an object is located at the near-point (the closest point at which an object can be brought into clear focus on the retina), the focal length of the cornea and the eye-lens must be changed so that the image is formed on the retina, which is still 20 mm away. If we take 25 cm as a typical value for the near-point, the algebraic form of the ray-tracing rules shows that the focal length of the cornea and the eye-lens must now be about 18.52 mm. The strength of the cornea and the eye-lens must now be about 1/0.01852 = 54 diopters. Thus the muscles of the eye must provide an accommodation range of 4 diopters. (This required range is actually independent of the size of the eye or the focal length of the eye lens when it is focused on a very distant object. It depends only on the value for the near-point.)

Myopia or "Near-sightedness"

In this condition, the strength of the cornea and the eye-lens combination is too great even when the muscles of the eye are completely relaxed. In other words, the focal length of the cornea and the eye-lens combination is always less than the distance to the retina, so that images of distant objects are formed in front of the retina even when the eye is totally relaxed. However, objects that are closer can be brought into focus. In many situations the focal length of the cornea and the eye-lens is so short that objects closer than the conventional near point of 25 cm can be brought into focus, which is why this condition is called "near-sightedness." People with this condition can thread needles and read books, but they often cannot read distant street signs.

Since the problem is that the strength of the eye-lens and the cornea combination is too great, the solution is to provide eye glasses (or contact lenses) with negative lenses. The negative lens weakens the strength of the cornea and the eye lens just enough so that the resulting focal length when the eye muscles are relaxed matches the distance back to the retina so that distant images are now in focus. Since the eye glass lenses are negative lenses, they are thinner in the middle than at the edges. It is easy to identify this kind of eye glass lenses, since acting by themselves they do not form a real image of an object at any distance.

Hyperopia or "Far-sightedness"

In this condition, the strength of the cornea and the eye-lens combination is too weak when the eye muscles are totally relaxed, so that the image of a distant object is formed behind the retina. It is usually possible to bring distant objects into clear focus by using the eye muscles to shorten the focal length of the eye-lens, but this accommodation is usually not sufficient to bring nearby objects into clear focus. Therefore people with this condition can focus on distant objects just fine, but cannot focus on an object at the conventional near point. People with this condition may not be able to focus on a book unless it is quite far away from the eyes, and they usually cannot thread a needle or do other close-up work.

The solution is the opposite of the previous case. The eye glass lenses are positive lenses which strengthen the cornea and eye lens just enough so that the resulting focal length when the eye muscles are relaxed matches the distance back to the retina. Since these eye glass lenses are positive lenses, they are thicker in the middle than at the edges. Acting by themselves they can form a real image of a distant object at their focal point, so that it is easy to distinguish between eye glasses for myopia and hyperopia, since the latter will form a real image of a distant object while the former will not.

Astigmatism

Astigmatism is the situation in which the focal length for the cornea and the eye-lens for an object oriented in some direction is not the same as for another object located in a perpendicular direction. In other words, the eye cannot bring the vertical and horizontal lines in a "+" symbol in sharp focus at the same time. (The axes of differing focal lengths need not be exactly horizontal and vertical.) This condition is often combined with one of the two previous conditions, but it need not be – it can exist even if the cornea and the eye lens have the proper average focal length and range of accommodation.

Since the problem is that the cornea or the eye lens are not symmetrical, the solution is to use eye glasses whose lenses are not symmetrical in a complementary way. An extreme version of this is a cylinder lens – a shape that is circular in one direction where it acts as a normal lens and flat in the other direction where it has no focusing properties. The strength and direction of the cylinder are designed to compensate for the asymmetry of the cornea or the eye lens. The curved portion of the cylinder lens may be positive or negative, depending on the detailed asymmetry in the eye.

It is easy to recognize eye glasses designed to deal with astigmatism, since acting by themselves they produce distorted images of any object. This is because they have different focal lengths (and therefore different magnifications) in different directions. Although people with mild hyperopia or myopia can often function without wearing glasses, people with astigmatism usually must wear glasses all the time. Otherwise they often get eye strain headaches caused by the eye muscles trying (and failing) to focus on the vertical and horizontal portions of an object at the same time.

The cylinder lens may be combined with an additional positive or negative lens if the person also has hyperopia or myopia.

Decreased accommodation or "Presbyopia"

This condition is typical of middle age. The eye muscles gradually weaken with age, so that the range of accommodation decreases. People with this condition cannot bring both near objects and far objects into focus. The weakening of the eye muscles often causes the focal length of the eye lens to increase as well, so that many people of middle age tend to become far sighted.

Since the problem is inadequate accommodation, no single lens can correct it, and people with this problem usually need bifocals, which are glasses with two different lens strengths: one for near and one for distant objects. The usual arrangement is that the bottom half of the lens is the near strength and the top half is the far strength.

Cataracts

This condition is also common among older people, although it can happen at any age. The lens of the eye becomes clouded and partially opaque. The result is that objects at all distances become blurred. The cloudiness of the lens often scatters light in all directions, so that people with cataracts are more easily blinded by bright lights. The lens often becomes yellow as well, and tends to absorb blue light.

The solution is to replace the lens with a fixed-focus lens of the correct strength. Since the replacement lens cannot be deformed, the result is that people with replacement lenses have no accommodation at all and must wear bifocals (or even trifocals).

THE SIMPLE MAGNIFYING GLASS

In this section we describe how a single positive lens can be used as a simple magnifying glass. Recall that the magnification of a lens is the ratio of the size of the image, s_i to the size of the object, s_o.The magnification of a simple lens is the ratio of the distance from the lens to the image, d_i divided by the distance from the lens to the object, d_o:

$$m = -\frac{s_i}{s_o} = \frac{d_i}{d_o}$$

The distance, d_i, between the eye-lens and the retina is fixed by the size and shape of the eye. Therefore, the magnification can be changed only by changing the distance between the object and the eye, d_o. Since this distance is in the denominator of the fraction above, the magnification increases as an object is brought closer to the eye. The maximum magnification will be realised when the object is brought to the near point. The magnification could be increased by bringing the object still closer, but the strength of the eye-lens can not be made great enough to bring the object into focus at these short distances. Since the distance between the eye-lens and the retina is about 20 mm and the near-point is about 25 cm, the maximum magnification of the eye is about 2/25= 0.08. (Note that this magnification describes the relationship between physical size of the image on the retina and the physical size of the object. All animals learn to relate the physical size of an image on the retina back to the physical size of the object that produced it.)

If a positive lens is placed in front of the eye, the combination of this lens and the eye itself result in a lens whose strength is the sum of the strengths of the two lenses.In other words, the focal length of the combination is smaller than the focal length of the eye itself. This compound lens produces an image which is in front of the retina and is therefore out of focus. However, since the image is now in front of the retina, the image can be moved back onto the retina by bringing the object still closer to the eye. This will bring the object back into focus on the retina and simultaneously increase the magnification. Thus the simple magnifying glass works by allowing the object to be brought closer to the eye while still being kept in focus.

Since the focal length of the eye-lens can be changed by the muscles in the eye, a calculation of the magnification of the additional lens in front of the eye depends on how the eye-lens is configured. These results are easy to calculate using the algebraic form of the ray-tracing rules.

There are two extreme situations:

1. The eye lens is configured with as short a focal length as possible so that it is still focused on the object at

the near-point. If a person then puts an additional positive lens whose focal length is f (in cm) in front of the eye, and if the eye-lens remains set to its shortest possible focal length, then the magnification (relative to the magnification of the eye by itself as described above) is

$$m = 1 + \frac{25}{f}$$

In order to be in sharp focus in this situation, the object must be placed at a distance of

$$d_0 = \frac{25}{25 + f} f$$

Since the denominator of the fraction is always greater than the numerator, the value of the fraction is always less than 1, so that the object distance is less than the focal length of the positive lens that has been added. For example, if the focal length of the additional lens is 5 cm, the magnification is 1+5=6, and the object will be in focus when it is about 4.2 cm in front of the eye.

The second situation is when the eye lens is configured to be as relaxed as possible so that it is focused on a very distant object. If a person then puts an additional positive lens whose focal length is f (in cm) in front of the eye, and if the eye-lens remains set to its longest possible focal length, then the magnification (relative to the magnification of the eye itself as described above) is

$$m = \frac{25}{f}$$

In order to be in sharp focus in this situation, the object must be located at a distance equal to the focal length of the additional lens. That is,

$$do = f$$

Since this distance is somewhat greater than the distance in the first situation, it should not be surprising that the magnification is somewhat smaller. Using the same additional lens as above, the magnification would be 5 and the object would be in clear focus when it was 5 cm in front of the eye

THE COMPOUND MICROSCOPE

The simple magnifying glass can magnify an image by a relatively small factor – usually no more than about 5. Although larger magnifications are possible, the aberrations of the lens become more important as the magnification increases. The magnifying glass also becomes more and more difficult to use, since the object must be moved closer and closer to the eye as the magnification increases.

The simple magnifier can be combined with other lenses in a number of configurations. Although all of the configurations are similar, each of these configurations is designed for a specific purpose, and the details are different.

The first such configuration is the compound microscope, which is designed to provide greater magnification than can be provided by a single lens. The microscope is designed to magnify objects that can be brought close to the device. The magnified image is inverted, but this is usually not a serious problem for most applications.

In its simplest form, the microscope consists of two positive lenses, which are mounted at either end of a tube as shown in the following diagram:

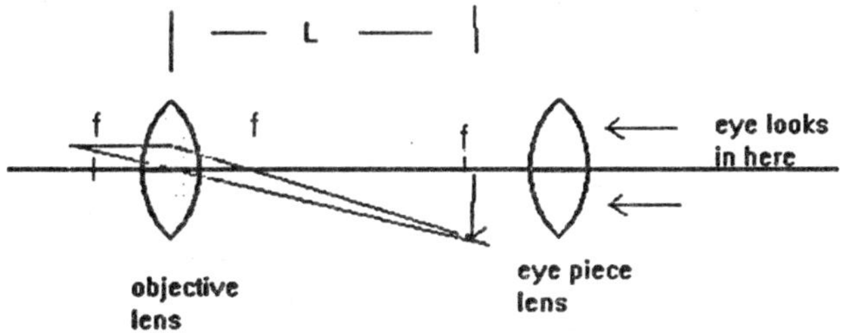

The first lens has a very short focal length, and the object is placed in front of this lens just beyond the focal distance. Since the lens is a positive lens, it forms a real, inverted image at a distance L behind the lens as shown. Since the distance

from the first lens to the image is much greater than from the object to the first lens, this image is magnified by the ratio of these two distances. The size of the length L is set by the geometry of the device, and is often about 16 cm.

For example, if the focal length of the first lens is 1 cm and if L is 16 cm, then, using the algebraic form of the ray-tracing rules, the object must be 1.07 cm in front of the lens (that is, just greater than the focal length of 1 cm). The magnification of the first lens is therefore 16/1.07 or about 15. More generally, the magnification of the first lens is approximately L/f_o, where f_o is the focal length of the objective lens in cm. The magnification could be increased by increasing L, but there is a limit to how big L can become before the microscope tube is too long to be used conveniently.

The second lens uses this real image as its object and acts as a standard simple magnifier. As we showed in the previous section, the magnification of this lens depends somewhat on whether the image it produces is viewed when the eye muscles are relaxed so that the eye lens is focused at infinity or contracted so that the eye lens is focused at the near point. In either case, the magnification is approximately $25/f_e$, where f_e is the focal length of the lens in the eye piece in cm.

The eye piece lens is often equipped with an adjustment that can move it in and out along the length of the axis of the instrument. This adjustment compensates for the different strengths of the eye lenses of different users. As we showed in the previous section, the combination of the lens in the eye piece of the microscope and the lens in the eye itself must produce a real image on the retina, and the eye lens in the microscope is adjusted until this image is in sharp focus there.

The overall magnification of the microscope is the product of the magnifications of the two lenses, and overall magnifications of 100X are relatively easy to obtain.

The magnification of each lens increases as its focal length is made shorter. Since the f/number is the ratio of the focal length to the diameter, decreasing the focal length also decreases the f/number. A smaller f/number means that the lenses capture more light from the object, which is a clear advantage. However, a smaller f/number also means that the depth of field will be correspondingly reduced. Since the amount of light is usually not a problem in a laboratory environment, the advantage of greater sensitivity is often not very important, whereas any decrease in the depth of field can be serious.

Therefore, the real trade off in the design of a microscope usually is the balance between magnification and depth of field. A common solution to this problem is to decrease both the focal length and the diameter of the objective lens so that the f/number, which is the ratio of these two quantities, remains more or less constant. Most high-magnification objective lenses are very small in diameter for this reason. The idea is to keep the focal length as short as possible to maximize the magnification while keeping the diameter as small as possible to increase the f/number and thereby give the user more depth of field.

THE ASTRONOMICAL TELESCOPE

The astronomical telescope is similar in design to the compound microscope. However, unlike the microscope, which is designed to view objects that are small and nearby, the telescope is designed to view objects that may be large, but are very far away.

As with the previous instruments, the magnification of the astronomical telescope is with respect to the performance of the unaided eye. If the distance between the eye lens and the retina is d, and if the object to be viewed is at a much greater distance D away, then the magnification of the eye lens itself is simply d/D, which is a very small number. The magnification of the telescope will be relative to this value.

The astronomical telescope consists of two lenses as shown in the following figure. The first lens, called the objective, has a focal length that is as long as possible. The focal distance of this lens is shown as "f" in the figure.

Since the object is a very great distance away, the first lens forms a real, inverted image just beyond its focal point. (The distance beyond the focal point of the objective lens is exaggerated in the figure above.) The magnification of the first lens is approximately f/D. Since D is a very large distance, this magnification is generally a very small number too. Nevertheless, although f/D may be a small number, it is already larger than the magnification of the eye lens by itself, which is only d/D, where d is the distance between the eye lens and the retina. The reason for this improvement is that the objective lens can have a focal length that is much greater than the focal length of the eye lens, which is roughly the diameter of the eye itself.

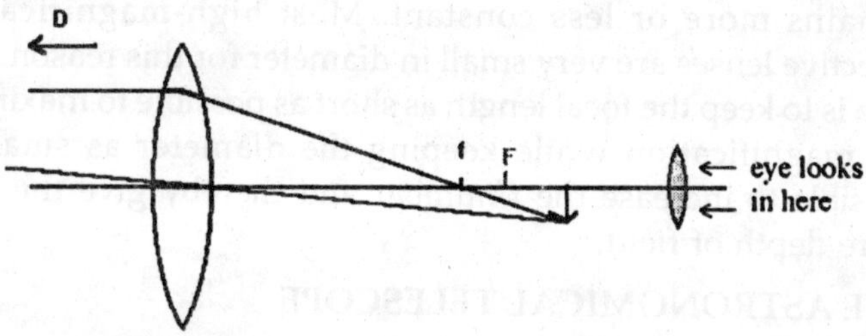

The real image formed by the objective lens acts as the object for the second lens, which is the usual simple magnifier. The second lens is positioned so that the image formed by the first objective lens is essentially at the focal distance of the second lens, which is shown as "F" in the figure. (The distances are exaggerated in the figure above.) Since the eye is positioned just behind the eye-lens of the telescope, the combination of the eye lens in the telescope and the lenses in the eye take an object that is essentially a distance F in front of the lens combination and produce an image on the retina a distance d behind the combination. The magnification of this combination is therefore d/F. The eye-lens in the telescope often has a small adjustment range to compensate for the different strengths of the eye lens and cornea in different people.

The magnification of the combination is the product of the two, or

$$m = \frac{d}{F} \times \frac{f}{D} = \frac{f}{F} \times \frac{d}{D}$$

The magnification is thus the product of two terms which can be re-arranged as shown. The first is the ratio of the focal lengths of the two lenses in the telescope and the second is the magnification of the eye lens acting by itself. The telescope therefore improves on the eye itself by the factor f/F. To maximize this value, the focal length of the first lens must be as long as possible and the focal length of the second lens must be as short as possible. The second criterion is the usual result for a simple magnifier.

From the figure, the separation between the lenses is the *sum* of the two focal lengths, so that increasing the focal length of the first lens increases the length of the instrument by the same factor. Thus a high-magnification version of this design tends to be a very long device - while decreasing the second focal distance can compensate to some extent, the second focal distance is usually so short already that its value does not make a significant contribution to the total length of the device. The length is sometimes made somewhat smaller by folding the length using prisms or mirrors. This is usually done in binoculars.

There is the usual trade-off between magnification, which increases as the focal length of the first lens is made longer and light-gathering power, which depends on the inverse of the f/number of the first lens and therefore decreases as the focal length is made longer so that the f/number is made greater. As with the microscope, many designs increase the diameter of the lens to keep the f/number at a reasonable value, but here the tendency is in the opposite direction from the microscope – that is, towards *increasing* the focal the length of the first lens (to increase the magnification) and simultaneously *increasing* its diameter (to keep the f/number and therefore the light-gathering ability at a reasonable value). The design criteria in the microscope tended to push both values in the opposite direction – towards smaller focal lengths and smaller diameters.

Note that the image produced by this telescope is inverted. This is usually not serious for astronomical observations, but a telescope designed for terrestrial use (binoculars, for example) usually have a second positive lens in the optical path that inverts the image again so that the observer sees an erect image of the distant object. This inversion can also be done with other optical elements (prisms, for example).

THE GALILEAN TELESCOPE

The Galilean telescope was first invented by Galileo in about 1609. Unlike the astronomical telescope, which is made using two positive lenses separated by the *sum* of their two focal distances, the Galilean telescope has a positive objective and a negative eye lens, which are separated by the *difference* between their two focal distances. As in the previous discussion, we consider an object at a very great distance, D, from the eye. The magnification of the eye itself is d/D, where d is the distance between the eye-lens and the retina.

The objective lens, whose focal length is f, performs the same function as in an astronomical telescope: acting alone, it would form a real, inverted image of the distant object, as shown in the following diagram:

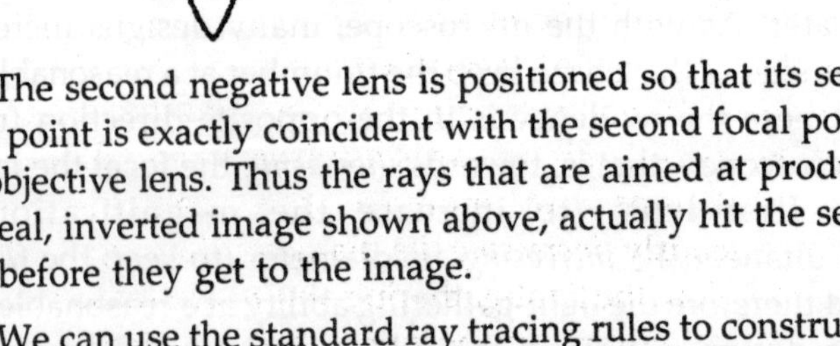

The second negative lens is positioned so that its second focal point is exactly coincident with the second focal point of the objective lens. Thus the rays that are aimed at producing the real, inverted image shown above, actually hit the second lens before they get to the image.

We can use the standard ray tracing rules to construct the image produced by the second lens, keeping in mind that the top ray in the figure above, which was headed for the second focal point of the first lens, is also headed for the same focal point of the second lens, since the two are coincident by construction. Using the ray-tracing rules for a negative lens, this ray is refracted parallel to the axis, and this is shown in red.

The dark green ray is another ray that forms the image produced by the first lens. Its direction is known from the figure above, since it originates at the object, is refracted by the first lens and ends at the image of the first lens. Note that it is also drawn to pass through the centre of the second negative lens. The second lens has no effect on this ray, since it passes through its centre.

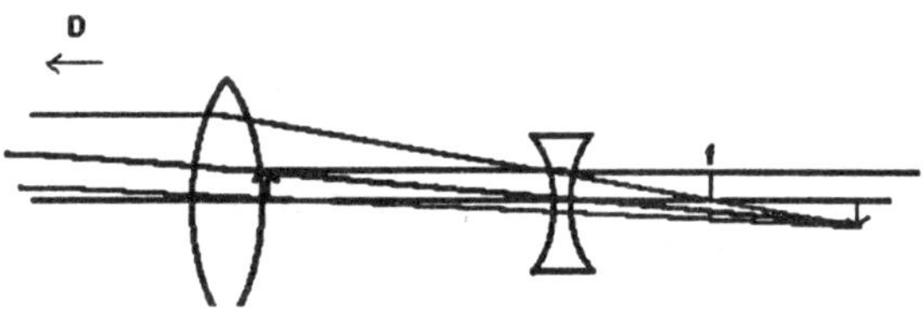

The dark green ray and the red ray appear to come from the purple image as shown. Therefore, the effect of the telescope is to produce a virtual, erect image of the distant object. Note that the image is erect rather than inverted. This virtual image is then viewed by the lens in the eye to form a real image on the retina.

Using the algebraic form of the ray-tracing rules, we can show that the magnification of this arrangement (relative to the unaided eye) is the same as for an astronomical telescope whose lenses have the same numerical focal lengths.

This arrangement is smaller than an astronomical telescope of the same magnification, since the lenses are closer together. In addition, the image is erect and so does not need a separate inversion as with the astronomical design. This configuration is often used in "opera glasses" and other simple binoculars where small size is an important requirement.

REFLECTING TELESCOPES

All of the refracting telescopes we have discussed, share a common limitation – the f/number of the telescope is limited by the size of the objective lens, and it is difficult and expensive to build high-quality lenses with large diameters. In addition, the weight of the lens increases at least as the square of its diameter, so that supporting a large lens at its edge is a hard job. Since the ray-tracing rules for lenses and mirrors are very similar, any optical design that can be realised with lenses can usually be realised with mirrors as well.

For example, an astronomical telescope constructed using two positive lenses that are separated by the sum of their two focal lengths can also be constructed using two concave mirrors with the same separation. This design is usually called a "Gregorian" configuration as shown in the following figure.

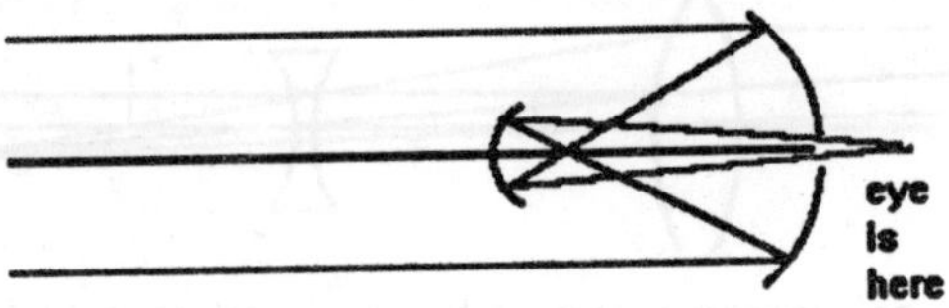

The basic idea is the same: the first mirror forms a real, inverted image near its focal point, and the rays from this image strike a second mirror where they are directed to the eye of the observer. Sometimes a simple magnifier lens is inserted into this path. Note that the second mirror partially blocks the field of view, but this is usually a small obstruction relative to the size of the main mirror. This configuration also requires drilling a small hole in the main mirror.

In addition to being easier to construct and support front-surface mirrors have no chromatic aberration, which is an advantage in many astronomical applications. Most large telescopes use mirrors rather than lenses for these two reasons.

In addition to the "Gregorian" design shown in the figure above, there are a number of other configurations, which differ primarily in the shape and location of the second mirror. Instead of using two concave mirrors as above, a "Cassegranian" design replaces the second small mirror with one that is convex. This concave-convex design is similar in concept to the Galilean telescope, which uses a positive-negative lens combination. There are also designs which use a combination of a large mirror to form the primary image and secondary lenses to focus this image into the eye or onto a sheet of film.

THE FIELD LENS

The ray tracing rules that we have discussed specify the size and position of an image in any position relative to a lens or mirror. However, these rules do not tell the whole story when the image formed by one lens is an object for a second lens. This situation arises in telescopes, compound microscopes, etc.

To illustrate the problem, consider the lens configuration of a compound microscope. The objective lens forms a real and inverted image of the object as shown in the following figure.

All of the rays from any point on the object are focused at the same point on the image. If there were a screen or some other target at the position of the image then all of these rays really would strike the target at that position, and the intensity of light striking any point would be given by the sum of the intensities of all of these rays.

However, if there is no physical target at the image plane, then while the rays still converge to a single point in the image plane, there is nothing there for them to hit and they continue traveling in straight lines.

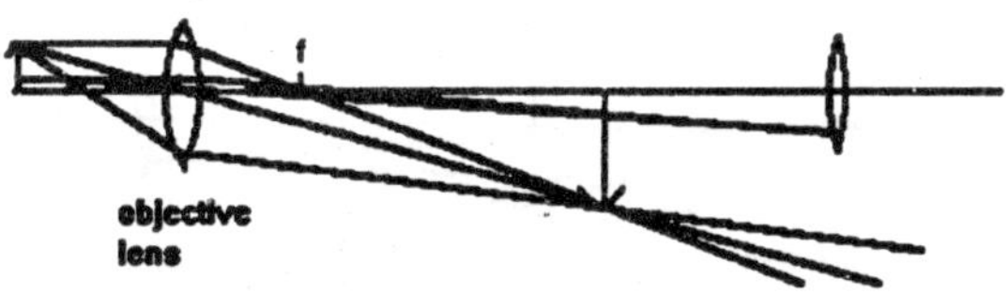

In the figure above, the red rays, which come from the top of the first object, converge to a point in the image plane as shown. However, if this is the image formed by the objective lens of a microscope or telescope, then there is no screen there. Instead, that image is the object for the second eye-piece lens. However, after converging to the point as shown, the red rays continue to travel in straight lines and never strike the second eye-piece lens. In other words, even though the objective lens formed an image of the tip of the original object, that point is not visible through the second lens since no rays from that point can ever strike it. Points on the original object closer to the axis can be seen through the second lens, as shown by the purple rays. This point on the original object can be seen through the eye-piece, since rays from the original object reach the eye-piece.

There are obviously intermediate points on the original object between the source of the purple rays, which has *all* of its rays striking the eye-piece lens, and the source of the red rays, which has *none* of the rays striking the eye-piece lens. For these intermediate points, *some* of the rays strike the eye-piece lens and so these points are visible but are not very bright. The bottom line is that the brightness of the image seen through the final eye-piece lens gradually gets dimmer for object points further and further from the axis, and there are points on the object which cannot be seen at all through the eye-piece lens.

The solution to this problem is to use a field lens, which is another positive lens located between the objective and the eye-piece as shown in the following figure.

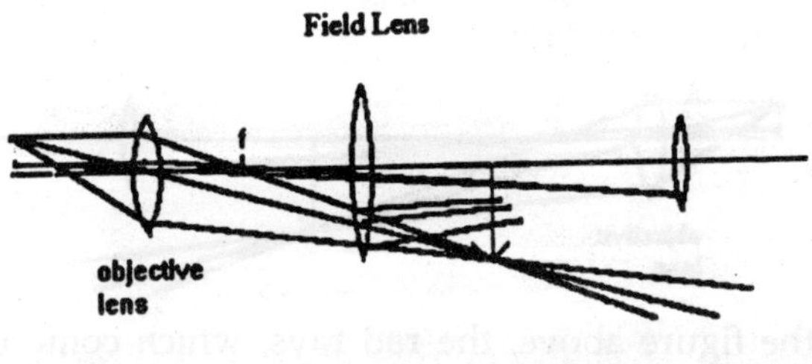

There are a number of different designs for the strength and position of the field lens, but the general idea is always the same. The field lens bends rays that would miss the eye-piece lens (or the next lens in a multiple lens system) back towards the axis, so that they pass through the next lens instead of continuing off into space. As an example, the effect of the field lens on the red rays.

On page 176, the text shows the field lens located exactly where the intermediate image would be formed if it were not present, but this is not always true. (If you are interested in this, look at the Hugyens and Ramsden designs for field lenses in a microscope).

COLOUR AND THE PRINCIPLE OF SUPERPOSITION

Many physical systems are assumed to obey the principle of superposition, which states that the response of some system to a combination of two inputs is the same as the sum of the responses to each input acting by itself. That is, Output(A+B)= Output(A) + Output(B). Bathroom scales, hi-fi amplifiers and many other common systems are assumed to follow this relationship. Such a system is also called *linear*.

Although it is not obvious from the definition, one of the important consequences of the principle of superposition is that the output of a system that obeys it does not contain frequencies that were not present at the input. This is especially important for hi-fi amplifiers, for example, where the output is intended to reproduce the input signal without additional spurious frequencies. For example, lower-quality amplifiers (which are not completely linear) can produce harmonics (integer multiples) of the input frequencies in the output signal. The output of such a system can also contain frequencies that are the sum or the difference of two frequencies that are present at the input.

We have already seen that the response of photographic film is logarithmic, and so photography using film does not satisfy the principle of superposition. It turns out that neither the eye nor the ear satisfy this principle either. The ear also has a logarithmic response to sounds, and the response of the eye is also non-linear in complicated ways. One of the consequences of these non-linear response functions is that both the eye and the ear can produce the sensation of frequencies which are not present at the input. Furthermore, the response of both of them to complex inputs having many frequencies is not given by the simple sum of the responses to each frequency individually.

Since light is an electromagnetic signal, there is a direct relationship between the frequency of a light signal and its wavelength, and the fact that the eye does not obey the principle of superposition with respect to frequencies also implies that its response to an input beam containing several wavelengths will not be the simple sum of the responses to each of the wavelengths individually. All of our characterization of colour is derived from this fact.

Hue, Saturation and Intensity

As we saw in the previous topic, when a beam of light containing many wavelengths strikes the eye, its response is not given by the sum of the responses that would be produced by each one of the component wavelengths acting alone. Although there are many different distributions of wavelengths that can be observed, our characterization of the appearances of these distributions can be expressed using a relatively small number of parameters. That is, there are many different combinations of wavelengths which appear to produce the same visual colour.

Although a number of different methods can be used to characterize a combination of wavelengths, it turns out that all of them use either 3 or 4 parameters. This is not an accident, since this small number of parameters is related to the way the eye perceives colour. The simplest triplet of parameters are called *hue, saturation* and *intensity*.

If we break up a beam of light into each of its component wavelengths and if we plot the intensity of each component as a function of wavelength, then, loosely speaking, the hue is the peak of this plot – the wavelength (or relatively small band of wavelengths) which has the greatest intensity. The hue is generally the single word that we would use to describe a composite colour. Hue values range from about 440 nm for violet, 450 nm for blue, up to about 700 nm for red light. The names associated with different hues follow the spectral decomposition of a rainbow: red, orange, yellow, green, blue, and violet. These descriptive colours are associated with ranges of wavelengths rather than with unique values, and some people can see colours outside of this conventional range of wavelengths (*ultra-violet* with a wavelength shorter than violet or *infra-red* with a wavelength longer than red).

The saturation of a beam of light is related to the width of the plot of intensity vs. wavelength described above. A completely saturated beam would have only one wavelength and would be called monochromatic, which a completely unsaturated beam would contain all wavelengths in equal proportion and would appear white. A completely saturated beam therefore has a very narrow intensity distribution function (possibly consisting of only one non-zero value in the limit), which a completely unsaturated beam has a very wide distribution function, possibly consisting of a constant value over most or all of the visible spectrum.

The intensity is related to the strength of the light beam. Intensity is very tricky to specify because the apparent brightness and the actual brightness can differ significantly. Loosely speaking, intensity is related to the total power in the light beam as measured by some objective instrument (such as a photographic light meter), but the perceived brightness of a light (or lightness of a surface) is strongly influenced by lots of other factors and cannot always be specified objectively.

These parameters are often not independent of each other. For example, the intensity and hue of a standard light bulb are related through the black-body relationships – decreasing the output intensity of a black body also shifts the hue towards longer wavelengths.

The hue, saturation and brightness of a light beam are often specified using a three-dimensional colour tree, as shown below. The vertical axis of the tree specifies the intensity of the beam, from nothing at the bottom (that is, black) through gray to some maximum value at the top corresponding to the brightest possible white. At each level of the tree (which corresponds to a given lightness or brightness), we draw a circle whose circumference shows the various pure, fully saturated, monochromatic colours of the rainbow in wavelength order from red to violet. The points on a radius line from the centre of the tree to some point on the circumference represent different unsaturated colours formed by mixing some amount of white from the centre of the tree with some amount of the colour at the end point of the line.

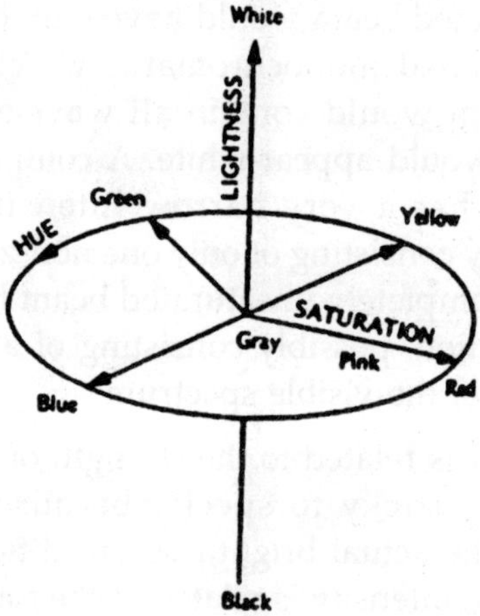

As shown in the figure, pink is a mixture of red and white, and the hue of the mixture is still red. Mixtures of yellow and white would produce colours that we would still call yellow, but which also might be described as "cream" white, "warm" white, etc.

ADDITION OF COLOURS

In the previous topic we talked about describing a light beam consisting of several (possibly many) wavelengths in terms of only three parameters: its hue, saturation and intensity. It is also possible to describe the colour of the beam by specifying the relative intensities of three colours which can be combined to produce a colour that looks the same (that is, has the same hue, saturation and intensity) as the colour of the original beam. This business can be complicated in practice, since the perceived colour of a light beam depends to some extent both on its intensity and on the colour of the surrounding objects. The fact that only three colours can be used to match the appearance of almost any other colour is not an accident – it is related to the way the eye perceives colour,

There are a number of different choices for the colours that can be used to match any other colour. It turns out that no triplet of colours can be used to match every possible colour. A good choice, which can match most colours, is the triplet 650 nm red, 530 nm green and 460 nm blue. (A somewhat "bluer" blue at 425 nm is sometimes used as an alternative to the 460 nm blue.) Using these three colours, it is possible to match many (but not all) other coloured beams.

If we combine these three primary colours in different combinations, we find that:

Red + Green = Yellow

Red + Blue = Purple (called *magenta*)

Green + Blue= Cyan (a dark blue)

Red + Green + Blue= White

Substituting each of the first three relationships into the last one, it is also true that

Yellow + Blue= White

Red + Cyan= White

Green + Purple= White

so that that these pairs of colours can be thought of as "complementary" – each one of the colours on the left side of these relationships can also be gotten by *removing* the other colour from a white beam.

Since any light beam of any colour can be described by giving the intensity of each of its constituent monochromatic frequencies, in principle we could reconstruct this arbitrary light beam by generating each of these frequencies with the proper intensity and combining them. Therefore, the process of reconstructing any arbitrary light beam can be completed successfully if we can use our three primary colours to match any monochromatic frequency. The job of matching the given light beam would then consist of adding the appropriate amount of each primary colour to match each of its constituent wavelengths taken one at a time. The following figure shows the relative percentages of the three primary colours needed to match any monochromatic colour in the visible spectrum.

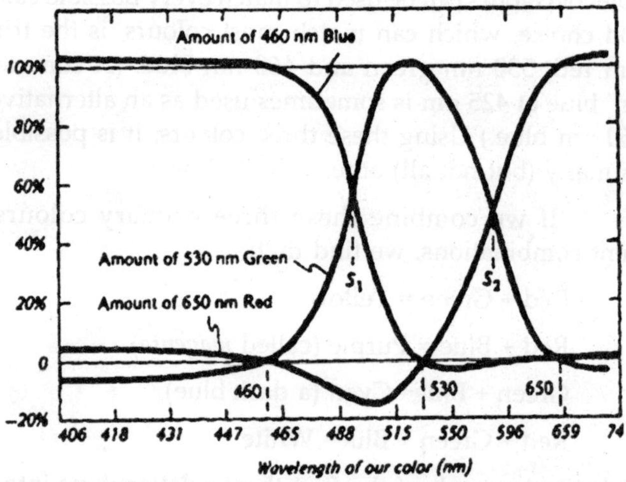

Note that there are some portions of the spectrum where a negative percentage of one primary is required, with another primary requiring a percentage of more than 100per cent — obviously both of these are impossible. For example, there is no way of matching a monochromatic blue beam that is bluer than the colour of our blue primary, which is 460 nm. We just don't have enough "blue-ness" to do the job, and the percentages are strange in that region as a result. The best that we can do is to add some green to the blue so as to decrease its saturation and then to match the combination. That is the significance of the fact that the percentage of green is shown as negative for these colours – that means that the green must be added to the colour we are trying to match.

The same sort of problem happens at about 500 nm, which we would call a yellow-green. We can't match a monochromatic colour in this region without adding some red to it. This red reduces the saturation of the initial colour, and we can then match it using a combination of blue and green as shown.

The following figure shows the same information in a different format.

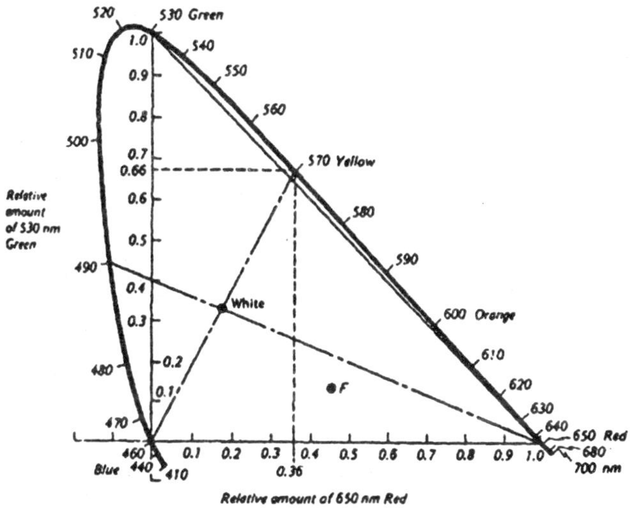

The x and y axes show the amount of 650 nm read and 530 nm green that must be added to match any colour. Since the percentages must always add up to 100per cent, any remainder between the sum of these two percentages and 100per cent is made up by our blue primary at 460 nm.

The horse-shoe shaped curve shows the visible wavelengths we are trying to match. The primaries that we have chosen are at the corners of the horse-shoe, which explains why they were chosen.

The triangle formed by the lines joining the primary colours represent the boundary range of colours that can be matched using these 3 primaries: colours inside of the triangular area can be matched exactly, while those outside of the triangle can be matched only by adding a "negative" contribution from one of the primaries – that is, by adding one of the primaries to the colour we are trying to match and then matching this combination using the other two primaries. As you can see from the figure, the largest region of difficulty comes in the blue-green portion of the spectrum, where the horseshoe curve is significantly to the left of the y axis, signifying that significant amounts of red must be added to any colour in this region of the spectrum in order for us to match it using a combination of green and blue.

In addition to specifying the combinations of the 3 primaries required to match any colour, this chart also shows what will happen when we combine *any* two other colours, even if they are not primaries. Thus the line joining the red and green primaries gives the colours that will result from mixing these two colours in various ratios. Likewise, the line joining 570 nm yellow (which is not one of our primaries) and 460 nm blue shows the various colours that will result from combining these two wavelengths in any combination. Since Yellow and Blue are complementary colours (that is, Yellow + Blue = White), the white point must lie somewhere on this line. The exact position of this line will depend on what we think of as our standard for yellow.

Using the chart above, we can see that combinations of red and blue would lie along the x-axis of the plot above. These combinations make various shades of what we call the colour "purple," but there is no wavelengths that correspond to these colours. Thus the various shades of purple cannot be produced using any single monochromatic wavelength. The colour purple therefore has a special, unique place in the spectrum – it can only be produced by adding red and blue in some proportion and cannot be made as a purely saturated colour. At least in principle, any other colour around the horseshoe could be produced in a monochromatic, 100per cent saturated form.

Since there is no unique wavelength associated with any purple colour, the best that we can do is to specify its wavelength in terms of its complement, which is some shade of green, which we can get by drawing a line from the purple through the white point to the other side of the horseshoe).

The "white point" – the point on the diagram corresponding to a completely unsaturated white is near (but not exactly at) the point at which the fractions of all of the contributing primaries would be the same (that is, 1/3, 1/3, 1/3). Part of this difference comes from the definitions of optical intensity, which include the variation of the response of the eye to monochromatic beams with the same physical intensity and part comes from exactly what we mean by a completely "white" colour.

From the definition above for the complement of a colour (that is, the colour that combines with this colour to form white), the complement of any colour is found from this diagram by drawing a line from the colour through the white point to the opposite side of the horseshoe.

Since the sides of the horseshoe above are curved, some colours cannot be matched by any three primaries, since a triangle joining any three primaries must lie inside of some part of the horseshoe. A triangle that was completely outside of the horseshoe could match any colour, but such a triangle would require using primaries that do not actually exist as real colours. There is a standard set of such imaginary primary colours which lie outside of the entire horseshoe curve and can therefore match *any* colour. These are called the tristimulus colours, and any colour can be matched using only positive contributions from these three imaginary primaries. The resulting fractions are called the chromaticity of a colour.

The principle of forming colours by adding these 3 primaries is used in colour television, in computer monitors and in similar applications. All of these systems share the principle deficiency of this method, namely that it is impossible to match all colours, especially those that are almost 100per cent saturated and therefore lie outside of the triangular lines joining the three primaries in the figure above. In addition to this fundamental limitation, real colour displays are limited by the fact that the materials that are used to display the various colours do not produce exactly the wavelengths specified above for the primary colours; the colours are also not completely saturated. These limitations are not noticeable for standard television displays, because the colours that are transmitted are not saturated anyway. Some of these limitations may be addressed in the newer transmission formats for "high-definition" television (HDTV), which will be available in a few years.

When colour television was first introduced, the format used for transmitting the colour information was chosen so that the signals could also be received and processed by black and white receivers. The method that was chosen to provide this compatibility was a variation of the hue, saturation, intensity format described above. In TV parlance, the black-and-white portion of the signal is called *luminance* and the colour portion is called *chrominance*. The luminance is roughly the same parameter as what we have called intensity, and the chrominance is a clever encoding of the relative amplitudes of the three primary colours that describe the colour at each point on the screen. The luminance signal is identical to the older black and white transmission format, so that a black and white receiver can process the luminance signal and recover a complete black and white version of the picture. The chrominance signal uses a transmission method that is not processed by the black and white system, and black and white televisions simply ignore this portion of the transmission.

Computer monitors generally employ much simpler encoding systems, since there is no need to transmit the information over a wireless channel or to remain compatible with previous standards. Many systems simply transmit the amplitudes of the 3 primary colours for each dot on the screen – in the older VGA standards each dot could have only 3 intensities: off, on/low, and on/high. This format gives 16 distinct colours for each dot (8 on/off combinations of the 3 colours and 2 overall intensities).

Finally, note that this description of colour does not tell the whole story, because the curves above specify the fraction of each of the 3 primary colours that matches some given colour. However, the colour that we perceive from a mixture of the primaries depends on the overall intensity as well as on the percentages of each component. For example, as we decrease the intensity of a light beam (keeping the fractional contributions the same), the colour "white" darkens and becomes various shades of gray. Likewise, orange darkens and becomes brown. This change in perceived hue does not affect the monochromatic colours – they look the same whether the beam is bright or dim. However, almost all of the unsaturated colours change hue as the intensity changes. To provide a complete description, the curves shown above would have to be repeated for every possible intensity.

SUBTRACTION OF COLOURS

The coloured appearance of most materials is due to selective absorption – the material absorbs some wavelengths very strongly and either transmits or reflects the remainder. The colour is therefore produced by subtraction – some wavelength or group of wavelengths is removed from the incident light. This is in contrast to producing colours by addition as described in the previous section.

If a light beam is processed by several such wavelength-selective absorptions, then what remains is the component of the light beam that is not absorbed by all of the processes. For example, if unsaturated white light is passed through a substance that transmits only the wavelength corresponding to the red primary (650 nm) then that is the only wavelength that emerges. If this beam is now passed through a substance that passes the green primary (530 nm) and absorbs everything else, then *nothing* comes out – the output is completely black. This is because the output of the first process is a wavelength that is absorbed by the second stage, leaving nothing in the output. This result will be true of any two substances that pass only a single monochromatic (fully saturated) wavelength and absorb everything else. The output of the first stage will consist of the single wavelength that is not absorbed. Unless the second stage passes exactly the same wavelength as the first stage, the output of the second stage will be black.

If the incident light is unsaturated white light, then the process of subtraction can be thought of using the ideas of complementary colours introduced in the previous section. For example, when unsaturated white light (such as the light produced by an incandescent light bulb) strikes a "yellow" object, the object appears yellow in this context because it absorbs the blue part of the incident light and reflects the yellow light. (Recall that in the previous section we identified this pair as complementary colours, since they combine to form white.) Continuing in this way, an object that absorbs the cyan from a white light will appear red and an object that absorbs the magenta from a white light will appear green. These colours: yellow, cyan and magenta, are called the subtractive primaries. They are defined as the answers to the question, "What colour must be absorbed from unsaturated white light to produce one of the 3 additive primaries red, green or blue. Using the colour horseshoe in the previous topic (or on page 245 of the textbook), the subtractive version of any colour can be found by drawing a straight line from the initial colour through the white point to the opposite site of the horseshoe, since the two colours at the opposite ends of this line will combine to form white.

Substances that generate the subtractive primaries function by absorption – they remove some part of the wavelength spectrum of the incident light so that what remains has some specified hue. Based on the definitions above, each one of the subtractive primary colours could be produced by passing unsaturated white light through an idea filter which completely absorbed one of the primary colours and passed everything else. Thus, yellow could be produced by passing white light through a filter which absorbed blue completely and passed everything else. Thus:

Yellow= Red + Green = White – Blue

Cyan= Green + Blue= White – Red

Magenta= Red + Blue= White - Green

If two such filters acted in series then the output is the part of the input that can pass through both filters. Thus passing unsaturated white light through a yellow filter absorbs the blue and lets everything else through. We could think of the result as a combination of only red and green, since the blue is gone. If the result is then passed through a cyan filter (which passes only blue and green) then the red will be absorbed leaving mostly green. Likewise, a magenta filter absorbs the green from white light, leaving red and blue. If the result is then passed through a cyan filter, the red will be absorbed and only the blue will be left.

The subtractive rules can be derived from these definitions and from the additive rules of the previous section. Thus,

Cyan + Yellow= (White – Red) + (White – Blue) = White – (Red + Blue)= (Red + Blue + Green) – (Red + Blue)= Green

Cyan + Magenta= (White – Red) + (White – Green)= White – (Red + Green) = Blue

Yellow + Magenta= (White – Blue) + (White – Green)= White – (Blue + Green) = Red

Cyan + Yellow + Magenta= (White – Red) + (White – Blue) + (White – Green)= White – (Red + Blue + Green)= Black

The left hand side of each relationship describes the combination of two filters acting in series on white light. In each case, the filter subtracts its colour from the incident light and passes everything else to the next filter in the sequence. The output of the combination is the only colour that is passed by both filters.

The subtraction process can be used to produce the same range of colours that can be produced by the addition process because the two are really different aspects of the same combination rules for complementary colours that we discussed in the previous topic. However, the subtraction process cannot create light when none is present to start with. For example, if, instead of using white light, monochromatic blue light is passed through a yellow filter, then the filter absorbs all of the input and nothing comes out – the output is black. The same effect is true for any filter and its complementary colour: if the filter is illuminated with only the complementary colour then the incident light is totally absorbed and the object looks black. In other words, the perceived colour of some object depends very much on the incident light. In particular, if the object reflects only colours that are not present in the incident beam then the object will appear black. In general, the perceived colour of the object will be given by the combination of the colours that it reflects *and* which are present in the incident beam.

In general, the colour that will pass through several subtractive filters is very hard to guess without a lot of detailed knowledge about the transmission characteristics. For example, a yellow filter might be constructed from a material that passes only 570 nm yellow and absorbs everything else. When white light is incident on this filter, everything but the 570 nm energy is absorbed so that the output light is a pure monochromatic yellow colour. If this light is passed through a sharp filter of any other colour, then nothing will come out and the filter will absorb all of the incident light and look black as a result. However, a yellow filter might also be constructed from a material that absorbs blue and passes red and green. When white light is incident on this filter, the filter absorbs the blue and reflects the red and green, which look like yellow. However, if *this* yellow is incident on a red filter (or on a material that reflects red), then the red component will be passed and the object will look red; if it is incident on a green filter (or on a material that reflects green) then the green component will be passed and the material will look green.

As with the additive mixtures discussed in the previous topic, the perceived colour from any mixture depends on the overall intensity as well as on the fractions of the constituents, so that these rules, which describe hue based only on fractions cannot tell the whole story. Varying the intensity of a light source is simple in principle, but complicated in practice. Although you can change the intensity of a standard light bulb by changing the input power (using a dimmer switch, for example), reducing the intensity in this way also decreases the temperature of the filament, which changes its hue towards the red portion of the spectrum. Changing the intensity in practice therefore means turning off some of the lights completely or partially blocking them using shutters or gray, "neutral density" filters, which absorb all wavelengths equally.

INKS, DYES AND PRINTING

The rules for forming colours by subtracting are important to understanding how colour printing, painting, and dyeing work. In all of these cases, the colour we see is formed by subtraction. Most inks and dyes are designed to reflect very little light, so that most of the incident light passes through the ink or the dye to the material below and is reflected back again to our eyes. The light passes through the absorbing material twice, and the colour that we see is determined by its absorption characteristics. (The reflective properties of the underlying substrate material are also important, of course, since the substrate material may also absorb some wavelengths are reflect others. To simplify matters, it is easier to think of the substrate material as reflecting whatever it receives without alteration and to treat the wavelength-dependence as coming from an additional layer on top of the simple white reflector.)

Color printing is generally implemented using multiple subtractive layers, where each layer is composed of one of the subtractive primaries defined in the previous topic. At each point, the colour of the light that we see is the sum of the wavelengths that make it through all of the layers at that point. Although it is possible to produce black using all 3 subtractive primaries in equal proportion, that takes a lot of ink, and most printing processes use a separate black ink for this purpose rather than using relatively thick layers of the 3 subtractive primaries. Color printing is therefore often called a 4-colour process for this reason, where the 4 colours are the subtractive primaries (cyan, magenta and yellow) plus black.

In addition to the fact that the inks are not ideal primaries, this method cannot be used to produce the monochromatic, fully saturated colours on the horseshoe colour chart itself. This is the same limitation that we saw in the process of colour addition. Also, recall that the subtraction process can work only with the light that is incident on it, so that it cannot produce a colour that is not present in the incident light beam. In addition, as the desired colour becomes more and more saturated, the resulting intensity inevitably becomes dimmer, since only a smaller and smaller fraction of the incident light matches the desired wavelength. Thus highly saturated colours do not print very well unless special inks are used. Since each one of these inks must be printed separately, the process is complicated and expensive, and is usually used only for "coffee-table" art books.

It is difficult to change the overall intensity of the reflected image, since printing is inherently a binary process – either the ink is deposited on a point or it is not. The usual solution is called a "half-tone" which is made by breaking up the image into a large number of dots. Each dot either receives the ink or does not. Since the dots are very small, the eye averages the intensity from many adjacent dots, so that the result is a perceived variation in intensity. This method can be used either with black ink (to produce a gray scale) or with coloured inks to vary the hue and saturation of the result.

PAINTING

Painting with water colours is very similar to printing with coloured inks as described in the previous topic. However, painting with oils is different and much more complicated. The reason for this complexity is that most oil paints reflect a significant fraction of the light that is incident on them, and the reflected light often has a different spectrum from the light that is reflected from the underlying substrate after having passed through the paint twice. These two beams combine again, so that what we see is a combination of the spectrum of the light reflected from the surface and the spectrum of the light reflected from the substrate. To make matters more complicated, the amplitude of the reflected light from a layer of paint depends on the difference between the index of refraction of the layer and the index of refraction of whatever is below it. Thus a layer of some paint may look different if is applied over another layer of paint instead of being applied over the bare underlying substrate (paper, canvas, ...). Since all paints reflect light to some extent, it is hard to produce a true black using the subtractive primaries, and a separate black paint is usually used – just as for printing.

To make matters worse, pigments in oil paints are usually made from readily available, cheap materials, so that they often do not coincide with the ideal subtractive primaries we have been discussing. For example, we recognize that the artist's simple rule, that Yellow + Blue = Green is not consistent with our descriptions. Yellow and Blue are complementary colours by our definition, so that the additive sum would be white and the subtractive sum would be black. Using our subtractive rules, we would have said that Yellow + Cyan= Green

While Cyan is certainly a shade of blue, it is not the monochromatic primary blue that we talked about above. The difference is that Cyan is a difficult colour to make in oil paints, so that the artist's rule is based on what is available in oil pigments, so that the artist's Green will not be quite the same as the green we have talked about. The same thing is true of magenta – it is hard to come by and expensive, so that it is not used as a subtractive primary in oil colours. Since red is easier to come by, both artists and printers often use "red" in place of what we would call "magenta." In fact, artist's red usually does have some blue as well, and it is not the same as the monochromatic red that we think of as a primary colour.

The pigment – the material that gives oil paint its colour, is usually a fine powder that has been suspended in an oil base. The base is often linseed oil, which is chosen because it is a polyunsaturated oil. Polyunsaturated oils are used as carriers of the pigment because they are liquids at room temperature but harden slowly when they are exposed to the oxygen of the air. The hardening produces a matrix of oil molecules that traps the powdered pigment. However, the reflectivity of the matrix changes as the oil cures so that the perceived colour also changes as the paint dries. Many oil based paints take a very long time to dry completely at room temperature because the surface of the paint cures first and prevents the lower layers from curing. In some cases, the painted surface must be heated to completely cure the paint.

House paints and paints used for other non-artistic purposes are often very different from the oil-based paints used by artists. They often use latex as a base rather than an unsaturated oil, and the curing process is quite different.

THE RESPONSE OF THE EYE TO COLOUR

The motivation for the previous discussions in which we tried to match any colour using various combinations of three primary colours is that this process is based on how the eye really perceives colour.

The rods of the eye – the sensors that are used when the light intensity is low, have some variation in their response to different colours. They are most sensitive to colours in the green portion of the spectrum near 500 nm, so that when the light is dim it is easiest to see objects that either radiate or reflect green light. However, since there is only one type of rod sensor, the rods cannot be used to discriminate among different colours, since they respond only to the weighted sum of the intensity that strikes them (where wavelengths in the vicinity of 500 nm green are more heavily weighted than those in the red or blue).

The cones of the eye – the sensors that are used when the light intensity is higher, come in three different types, where the differences among the types are the wavelength bands they respond to most strongly. The three types of cones respond to all wavelengths to some extent, so that every colour excites them all – at least to some extent. However, the amplitude of the response of each type is a function of the incident wavelength.

Unfortunately, each type of cone does not provide any wavelength discrimination beyond its overall response function. In other words, the short-wavelength cone, which is sensitive primarily to a broad band of colours in the blue and green portions of the spectrum, cannot provide any information on the wavelength of the incident light beyond the fact that it has been triggered by some wavelength within its response band. Color discrimination depends on the fact that every wavelength excites the three types of cones by different amounts – every incident wavelength produces a unique triplet of excitation values. This is an important design feature: if a band of wavelengths excited only one type of the cones, then we would not be able to distinguish differences in hue within this band, and if the response functions did not overlap, then there would be wavelengths that we would not be able to see at all, since those wavelengths would not excite any one of the cones or would produce so little excitation that these wavelengths would appear very dim. Finally, if the response functions of the cones were essentially identical over any range of wavelengths then it would be difficult or impossible to distinguish the hues in this wavelength range.

The details of the response functions of the 3 types of cones can be derived from the observed response of the eye to various colours – especially to the fact that only two complementary colours can combine to produce white. These pairs of colours must therefore straddle the cross-over points between the sensitivities of the 3 cones, so that the two colours can excite all 3 types of cones more or less equally so as to give the sensation of white.

Additional information comes from people who are "colour blind." These people often have only two of the 3 types of cones, so that every colour that they see is some combination of the two types of cones that are working. For example, if the green cones are not working, then these people discriminate among colours based on the red and blue content alone, and they cannot distinguish colours that have the same red and blue content but which differ only in how much green they contain. In other words, they see all colours as if they were on vertical lines through the colour horseshoe (figure 9.11, page 245), since colours along these lines differ only in how much green they contain.

The result of these considerations is that the response curves of the 3 types of cones are as shown in the following figure, which is taken from the textbook on page 273 (figure 10.5). The response function marked S responds most strongly to the shortest wavelengths: blue and violet; the response function marked I responds most strongly to yellow and green, while the response function marked L responds to the longer wavelengths, primarily yellow, orange and red.

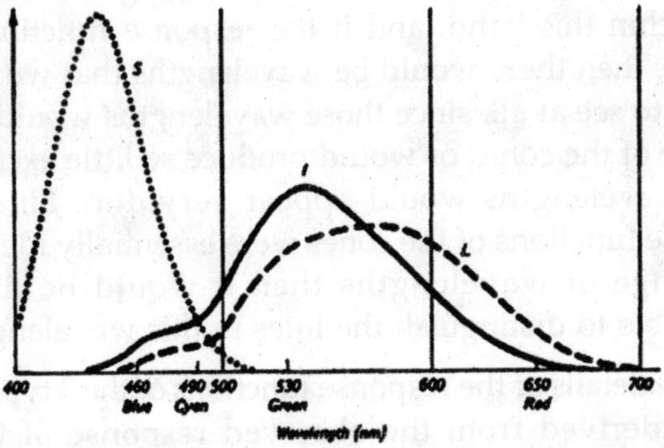

These response curves show how the blue and green primary colours were chosen, since they are close to the peaks in the response functions of two of the 3 types of cones. (Also note why some definitions use a primary blue at 425 nm – it is closer to the peak response of one of the types of cones than 460 nm is.) The 3rd primary has a longer wavelength than the peak in the response of the third cone, which is why monochromatic red colours tend to look less bright than the same amount of energy at shorter-wavelength parts of the spectrum. In addition, the response of the "red cone" to red light is only somewhat greater than the response of the "green cone" to the same wavelength, so that red light often has a hint of yellow in it even when it is really a completely saturated red.

The fact that there are three types of cones whose wavelength sensitivities are as shown above explains many of the phenomena that we have described. For example, it is clear why colour matching using the three primary colours can work. Since each type of cone reports only its relative excitation level and not the exact wavelength that caused the excitation, it is possible in principle to match the three excitation levels resulting from any incident wavelength using an appropriate combination of the three primaries.

It is also clear from the figure why some colours (especially monochromatic blues and greens) cannot be matched using the usual primaries. Any incident wavelength in the range of 460-500 nm will excite the blue cone quite strongly, but will also excite both the green and red cones to some extent. Thus matching colours in this region will need lots of the blue primary. Since the green cone is also excited, some green primary will also be required. Unfortunately, the response of the red cone to this green light is quite significant – much more than the response of the red cone to the incident wavelength. Thus adding green primary to the mix almost always produces too much excitation in the red cone. The only way to match this unintended (but unavoidable) excitation of the red cone is to add some red to the incident wavelength. This conclusion is exactly the same as we came to from considering the colour matching horseshoe chart in the previous topics – highly saturated blue/green wavelengths require a "negative" red contribution in order to be matched.

Although the response curves of the cones explains colour many of the aspects of colour matching, it does not explain all of the observed phenomena of colour vision. In particular, the fact that the combination of two colours (such as red and green) can produce a third colour, whose wavelength is not related in any obvious way to its two parents is not a direct result of the previous discussion. In addition, the interaction between intensity and perceived hue is not explained either. Since the physical response functions of the cones have been verified in many ways, these additional phenomena must be implemented in the processing of the signals from the cones rather than in the hardware of the cones themselves.

While there are only three types of cones, there is some evidence that the processing of these 3 signals can be characterized in terms of *differences* between the different channels. There are 5 signals in this model: red, green, blue, yellow and intensity. (Based on our previous discussion, the yellow channel is presumably implemented internally as (red + green). The 3 signals that are processed by the brain are the pair-wise differences: red – green, blue – yellow and the overall intensity, which is presumably estimated as the sum of the excitations of the 3 types of cones. The intensity response function is complicated by other factors. For example, the perceived brightness of an object depends on its surroundings, so that the intensity channel incorporates some kind of spatial differencing.

The wavelength values corresponding to these physiological primaries are not quite the same as the primary colours of the previous discussion. The largest discrepancy is the value of the "red" primary, which is closer to what we defined above as magenta – a mixture of red and blue. Therefore, using our previous primary values, the red – green signal is really more like (red + a little blue) – green. Since there is no cone that explicitly responds to yellow, the yellow – blue channel is presumably implemented as (red + green) – blue.

When both difference channels red – green and yellow – blue respond with a 0 value, then the colour is some shade of white depending on the value of the intensity channel. The overall response is summarized in the following figure.

In addition to the differencing in wavelength, there is also some spatial differencing. That is, a colour in one area can affect the perception of the colour of an adjacent area. That is, a bright yellow area makes an adjacent uncoloured area appear blue, etc.

Finally, there is an interaction between the perceived colour of a scene and its actual average intensity. This tends to work in two ways:

1. Since the colour processing involves differences between channels, a decrease in the overall illumination does not affect the differences as much,

so that we can perceive more or less the same colour under different levels of illumination.

2. A change in the intensity of only part of a scene relative to the rest of it does have an effect on the perceived hue – dim orange will look brown, for example.

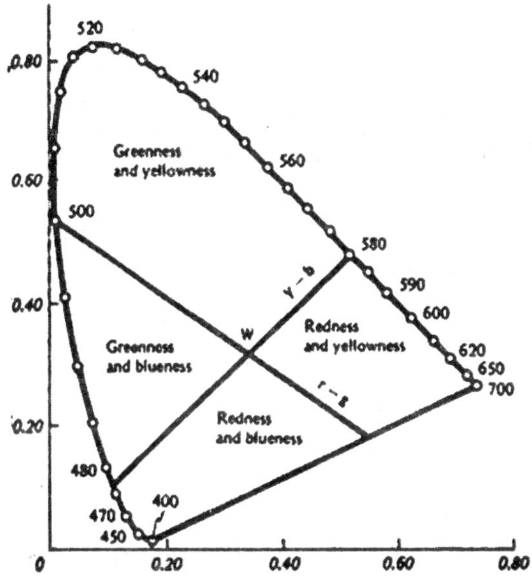

FILMS FOR BLACK AND WHITE PHOTOGRAPHY

We have already discussed the ASA and DIN methods for characterizing the sensitivity (or "speed") of photographic film. In this topic we describe some of the chemical processes that are involved in photography. Almost all films use light-sensitive chemicals based on silver halides that are coated on an inert base material such as glass or plastic. (A silver halide is a compound that includes silver combined with one of the four common halogens: fluorine, chlorine, bromine or iodine. Silver fluoride has very little light sensitivity, and many films use either silver chloride or silver bromide.) Most modern films use coatings based on gelatin emulsions: relatively viscous substances with tiny grains of the silver halide dispersed throughout the material. (Think of something like mustard.) The size of the silver grains varies somewhat among different films, but they are typically a few microns in diameter (1 micron= 10^{-6} m= 0.000 04 inch). Given this size, there are something like 10 million grains in a typical 35 mm negative. Silver halides by themselves are not very sensitive to the longer wavelength portion of the spectrum (reds and yellows), and the original halide-only films were sensitive only to blue. The emulsion in modern films also contains additional chemicals to make the film sensitive to the longer wavelengths as well. This type of film is called panchromatic (sensitive to *all* colours). All general-purpose black and white film is of this type. The older orthochromatic films, which are sensitive to blue and green but not to red, are still used for special applications. Even some panchromatic films may be somewhat more sensitive to blue than to red, so that objects with lots of blue in them (like the sky) appear to be too bright. For this reason, many photographers use orange or yellow filters in front of the lens when using black and white film. These filters remove some of the blue, and balance the exposure of the blue sky and the non-blue ground.

All of the silver halides are unstable to some degree. They will decompose spontaneously liberating free silver, so that any film that uses these chemicals has a finite shelf life. Cooling the film decreases the rate of spontaneous decomposition, so that many photographers keep film in a refrigerator or freezer until it is used. The silver halide molecules are decomposed when the molecules are struck by light, liberating free silver atoms. The number of grains which have atoms of silver formed in this way is proportional to the intensity of the incident light. (Recall that the proportionality is logarithmic rather than linear.) In general, only a relatively small number of atoms of free silver are formed in this way, so that there is no visible image on the film at this point. Since the incident light is divided among all of the grains, there is an inverse relationship between grain size and sensitivity (as measured by the ASA number, for example). Other things being equal, a film with smaller grains will have greater resolution (because the resolution in the image is directly related to the size of the grains) but a lower ASA value, since the intensity striking any one grain will be less.

The film is "developed" by placing it in a solution that would eventually decompose all of the halide molecules, liberating free silver. The rate at which this decomposition proceeds depends on how many free silver atoms are already present in each grain, so that the developer tends to produce more free silver at those locations where the exposure to light had already started the process. If the development process is stopped at the proper moment, only the silver halide in those grains that initially had atoms of free silver (as a result of having been exposed to light) is converted to metallic silver. Grains that had no free silver initially are not altered. If the developer and the residual unaltered silver halide are removed, the result is a "negative" silver image. This negative image has lots of opaque free silver where the incident light was strong, some free silver at points where the light was weaker and only the transparent backing material where there was no incident light to begin with. The negative is "printed" by shining a light through it onto a second piece of photographic paper. The development process is repeated, the image is inverted again, and a positive image results. The gray scale both in the negative and in the final print is produced by varying the amount of opaque free silver.

"Reversal" films can produce a positive image in one step. These films are used for preparing slides and transparencies for projection. The reversal to produce the positive image is accomplished as the film is developed. The process of developing the film starts the same way as for a negative image described above. After the developer has converted the latent image to one containing free silver at each point where light struck the film initially, the silver is washed away (the process is often called "bleaching") leaving only unmodified silver halide where there was no light initially. This remaining unmodified silver halide is exposed to a second flash of light.

The process of developing the film is repeated, and now free silver is formed at these locations where there was no light initially. The development process is stopped at the appropriate point and the remaining steps of washing and drying are the same as above. When this film is placed in front of a projection lamp, the free silver formed during the second exposure blocks the light and produces a dark point on the screen. The points which had free silver formed during the first exposure are transparent because that silver has been washed away. The result is a positive image.

Black and white reversal films exist, but they are not very common any more. They were used to produce slides of black and white subjects – the kinds of things that are now done by electrostatic copying onto transparent plastic sheets which can be displayed by view-graph machines. The same sort of reversal process is used to produce colour slides, as we will discuss in the next topic.

FILMS FOR COLOUR PHOTOGRAPHY

Color photography is based on the two topics we have already discussed: forming black and white images on film and matching any given colour using the three primary colours. In essence, colour photography consists of forming three independent images using the light reflected from the object in the 3 primary colours and then combining those 3 independent images to reconstruct the initial colour of the object. Each one of the three images is formed using techniques very similar to those that are used for black and white photography as described in the previous topic. The technical tricks involve how the 3 separate red, blue and green images are produced initially and how they are recombined in the final print. Since all colour photography is based on using the three primary colours (either by addition or subtraction), it is affected by the same limitations as other processes that use this technique. That is, it is impossible to reproduce colours that are very saturated. Different brands of film use different dyes and slightly different developers, but all of them are basically the same and are affected by this limitation.

As with black and white photography, colour photography is divided into two broad categories –

1. Color-print films, which produce a negative image after one development and a positive image after a second exposure in the development process, and
2. Reversal films, which produce a positive image on the film itself after a more complicated development process. Reversal films are used for slides and movies.

The early colour films were usually reversal films for slides or movies, and were based on forming colours by addition of the 3 primaries: red, blue and green. The light from the object being photographed was divided into 3 beams using red, blue and green filters, and the 3 beams were used to produce 3 images. One of the early techniques for achieving this separation used standard black and white film with a distribution of tiny colour filters spread more or less uniformly over the entire film area. (The filters were added to the emulsion more or less randomly across the negative during the manufacturing process.) Each filter passed light only in one of the three primary additive colours to the film below. The density of silver below each filter corresponded to the intensity of light in that primary colour at that point. The film was then developed using the standard reversal technique, which produces a clear spot on the film wherever the initial exposure contained lots of free silver and a dark spot on the film wherever the initial exposure contained no free silver. When the film was viewed through white light (in a slide projector, for example), the combination of the transparency of the film (which was proportional to the intensity of light from the object at that point) and the tiny filters in front of each point on the film re-created a full colour image by the usual method of 3-colour addition. (For example, a clear spot underneath a red filter implied that the original object had a lot of red light at that point. When the developed film was projected, the red filter in front of that clear spot passed red light to the screen. Conversely, a dark spot on the film meant that the original object had no red light at that point, because that dark spot was created during the second development. Thus no red light was transmitted at that point. Each of the other filters worked the same way.)

A variation of this process used three physically distinct films, each of which was exposed to the light from the object in only one of the primary colours. Each of the films was then processed to produce a reversal image, and the three films were then projected onto a screen using three independent projectors, each of which was equipped with the proper filter, which only allowed light of one primary colour to reach the screen. This method is pretty cumbersome for films, but is exactly the method used in some projection-TV systems.

The primary advantage of systems based on the additive primary colours is that the film itself is standard black and white film, and can be processed by the usual developers to produce a reversal image. This was a significant advantage in the early days of photography, when most developing was done by hand, because colour developers are tricky to use and require very careful control of the development time and temperature.

One of the limitations of this class of methods was that it was difficult to make copies of the picture without degrading its sharpness. This degradation resulted from the fact that since the tiny filters were arranged more or less at random across the film surface, different pieces of film had the 3 filters arranged in different semi-random ways. A point that was bright red on one negative might not be copied properly because the film used for the copy did not happen to have a red filter at exactly the same point. The fact that the images in the 3 primary colours are separated in space to some extent is a fundamental limitation of all of these methods, since the resolution of the image is limited by this spatial separation in a fundamental way. (Current TV pictures suffer from the same limitation.) The obvious solution to this problem is to use a process based on subtractive primaries, since the three images can be exactly coincident if subtractive primaries are used.

Most newer films differ in two important respects from the earlier films based on the additive primaries. First of all, they work by colour subtraction rather than by addition, and second of all, the different colours are generated using dyes – transparent materials (similar to the inks used for printing) that function by subtraction by absorbing one of the positive primaries. As we mentioned in the previous paragraph, colour subtraction is capable of higher resolution in principle, since the three colour images can be superimposed on each other to form the final full-colour image.

The earliest colour films based on subtractive primaries used three physically different pieces of film for each of the three colours. Each film recorded the image in one of the primary colours. Each of the three films was processed so that the thickness of the emulsion at each point of the negative was proportional to the intensity of the corresponding primary colour that struck the film there. For example, if a particular point on the object had a lot of blue light, the emulsion of the film that records the blue intensity was thick at that point. Each film was then soaked in a dye that was the complement of its primary colour, and the amount of dye that was absorbed was proportional to the thickness of the emulsion. The point which had a lot of blue, for example had a correspondingly thick emulsion in the film that recorded the blue primary colour, and that emulsion absorbed lots of yellow dye. The three films could then be combined using the usual reversal process to produce a positive full-colour image. This idea is the basis for the "Technicolour" process. This process is too complicated for everyday use, since it requires three separate cameras and a complicated processing scheme. Although later versions of the Technicolour scheme used only one camera, the system was still too cumbersome for most non-professional users.

Instead of using the thickness of the emulsion to control the contribution of one of the primary colours, an emulsion for a colour film consists of silver halide with a chemical called a "coupler." The coupler is a chemical that reacts with the developer to form a coloured dye wherever free silver is formed. There are two general techniques for introducing the couplers into the film. A *substantive* dye coupler is introduced into the emulsion layer during the manufacturing process. This idea is used in Ektachrome film. A *non-substantive* dye coupler is introduced during development. Kodachrome and Fujicolour are based on this technique. There are advantages to either technique.

Instead of a single emulsion of gelatin and silver halide, a colour film has three separate emulsions — one for each of the three primary colours. The top emulsion is a standard pure silver halide, which is sensitive only to blue light. In a substantive colour film, this layer also contains a chemical that produces a yellow dye (the complementary colour to blue) during the development process wherever there is also free silver. Thus the density of the yellow dye produced in this layer at any point is proportional to the intensity of blue light from the object that struck that point originally. In a non-substantive film, the yellow dye that winds up in this layer is introduced during the development process, as described below.

Since the following layers are also sensitive to blue light (in addition to the another primary colour) the blue emulsion is followed by a yellow filter, which absorbs the blue but passes red and green. The next layer is usually green sensitive – as above, in a substantive film, the silver halide is combined with a coupler that produces a magenta coloured dye, which is the complement of green. A non-substantive film has the magenta dye for this layer inserted during the development.

The final layer is the red sensitive layer, which contains the usual silver halide combined with an appropriate sensitizing chemical to make it red sensitive. This layer is combined with a cyan coupler (either during manufacture or development as above).

The developing process for a negative substantive film is similar to the process used for black and white film, but the chemicals are different. The developer acts on all of the layers simultaneously. It produces silver grains in each layer wherever light of the corresponding colour was present in the original object. The silver combines with the coupler and the developer to produce a spot of the dye at that point. The colour of the dye is the complement of the original colour, so that the resulting negative is reversed both with respect to colour and with respect to intensity. That is, a point where lots of red light struck the film will have lots of free silver in the bottom emulsion layer, which is red-sensitive. During the development process, the free silver combines with the coupler to produce cyan dye at that image point (the complement of red). (A cyan dye looks cyan in colour because it passes cyan and absorbs red – just like the cyan inks used in four-colour printing.) A point which had no red light from the object will have no free silver and no cyan dye. The unexposed silver halide and the silver produced during the development are washed away, leaving just the dyes formed in each layer. Points where no dye was formed are transparent in that layer. Thus if some point on the object produced only red light, the corresponding points on the blue and green layers of the film are clear, while the red layer has cyan dye at that point (which absorbs red).

A positive print is produced using a second exposure as for black and white prints. A white light is placed behind the negative, and the image is projected onto a second piece of film or photographic-sensitive paper. The point on the negative which has cyan dye as a result of the first development process passes cyan and blocks red. Since the other layers at that point are clear, the blue and green components of the light pass through the negative without attenuation. The light that passes through the negative at that point therefore contains all colours but red. When this light strikes the second photosensitive paper (which will be the final print), the silver halide in the blue and green layers are converted to silver, producing yellow and magenta dyes at that point; the red layer of the second exposure is not altered since the cyan dye in the negative removed the red light. The unexposed silver halide at that point will therefore be washed away at the end of the second development, leaving a clear spot in that layer. When the resulting image is viewed in white light, the yellow dye in the first layer absorbs blue, the magenta dye in the second layer absorbs the green, and red light is not absorbed by the clear third layer. The point looks red – just like the original point on the object. Conversely, a point that had no red light in the original object has no cyan dye in the negative and that point on the negative is clear. During the second exposure, the red light converts the silver halide in the red layer to silver, which combines with the coupler to produce cyan dye at that point. When that point is viewed in white light, the cyan dye absorbs the red component at that point, so that the reflected colour has no red it. Thus a red point on the original object results in a point that appears red in the final image and a point that had no red in the original object has cyan dye at that point and does not appear red.

The same thing is going on simultaneously in the other layers. To summarize, after the first development, the negative consists of three layers of dyes, which are the complements of the original image both with respect to intensity and colour. Bright parts of the original object have lots of dye in all three layers and appear black because of the usual 3-colour subtraction rules. The same is true for all other intensities and colours. The second exposure and development of the printing process reverses the whole thing again and produces a positive image. Note that unlike printing, there is nothing that corresponds to the 4th black ink that printers use, so that we would expect that subtractive colours films would have difficulty reproducing an object that had lots of black in it. (Fortunately, such objects are not very common, but this fact explains why colour film is not used to produce slides which have mostly black and white content, such as slides containing only text.)

The development process for a film based on the non-substantive process is similar, but the details are different. Since the film does not contain the dye couplers, these must be introduced during the development process. There are usually three phases to the development process – each of which is used to convert the latent image in one of the layers to free silver and introduce the appropriate dye coupler at the same time.

A colour slide is produced using the analog of the reversal process described for black and white reversal positives. The first development converts the latent image into free silver and the dye of the complementary colour. These products are then washed away and the film is exposed to light a second time. The areas that were not exposed to light from the initial object are now exposed, the remaining silver halide is converted to silver, and the silver interacts with the dye to produce the coloured image. For example, a point on the object that produces only blue light converts the corresponding point on the blue layer of the film to silver and yellow dye during the first development. The remaining layers are not altered. The yellow dye and the silver are washed away leaving a clear substrate on all of the layers at that point.

The film is then exposed again to white light. The red and green layers are now affected, producing cyan and magenta dye at that point, respectively. The blue layer is clear at that point, so nothing happens there. When the film is placed in a projector, the cyan and magenta dyes absorb the red and green light, respectively, and the blue portion of the light passes through the film and is seen on the screen.

A point on the object that had no blue light does not affect the corresponding point on the film. The halide in the blue layer at this point remains unaltered until the second exposure converts it to silver and yellow dye. The silver is washed away leaving yellow dye at that point. When the film is projected, the yellow dye at that point absorbs blue, so that the projected image of that point contains no blue light, which is exactly the same as the original object. The other layers behave in the same way, so that the resulting image has the same colour balance as the original object.

The sensitivities of the three layers in colour films are adjusted to match the kind of light that will be used to illuminate the subject. For example, daylight is much bluer than incandescent light bulbs, so that a film designed to be used outdoors has its sensitivity to blue light reduced relative to the sensitivities at red and green. If this film is used to photograph a subject illuminated by normal incandescent lights, the resulting picture will look too yellow, because the reduced sensitivity to blue built into the film combines with the relatively small amount of blue light from the light bulbs to produce a picture that is lacking in blue and therefore looks yellow.

Conversely, an "indoor" film designed for incandescent lights has its blue sensitivity enhanced to compensate for the relatively small amount of blue light emitted by incandescent light bulbs. It will produce photographs that are too blue if the subject is outdoors. Finally, although fluorescent lights appear to be white by eye, they contain lots of green and yellow (because of the very strong lines in the mercury spectrum at these wavelengths), and these strong colours will affect the colour balance of all colour photographs.

Black and white films also have a sensitivity that varies with the colour of the incident light. Most black and white films are more sensitive to blue light than to light of any other colour, so that objects that have lots of blue in them (like the sky) appear much brighter in a photograph than equally bright objects of other colours. Some black and white photographers use a filter in front of the lens to correct this imbalance. The filter is designed to absorb sky blue, and therefore has a yellow/orange colour.

The dynamic range of a film specifies the range of light intensities that the film can handle. Although this value varies from film to film, most black and white films have a dynamic range of about a factor of 100 (that is, almost 8 f/stops). In other words, the brightest point on the image can be about 100 times brighter than the dimmest point. All points brighter than the brightest point cannot be distinguished – they all look white, and points dimmer than the dimmest point all look black. This dynamic range is a function of the film, but it can be moved around with respect to the brightness range of a complex object to overexpose or underexpose some part of the object. Color films have a somewhat smaller dynamic range because the exposure interacts both with the brightness of the image and with its colour balance. The dynamic range is usually about a factor of 50 (that is, 5 or 6 f/stops).

Chapter 10

Interaction of Light with Matter

INTRODUCTION

Now that we have some understanding of the energy level structure of atoms and molecules, we need to concern ourselves with how these interact with electromagnetic waves. Such interactions are at the very heart of spectroscopy. There are lots of spectroscopic processes. The important ones are:

- Refraction
- Rayleigh scattering
- Brillouin scattering
- Raman scattering
- resonance Raman
- Absorption
- Resonance fluorescence
- Fluorescence

They appear to be different but in fact they are all closely related. We would like to examine these in much greater depth to understand them and their relationships.

A SIMPLE ANALOGY

The most naive way to discuss absorption and emission of light is to consider the atom and its electrons as electrons on springs. You might object and say that is a ridiculous comparison and it is. However, the mathematics are identical and a great deal of insight can come from the analogy. If you have a weight on a spring, you can set it into motion by tapping it at its natural resonance frequency. The best way to input energy to it is to tap it 90 degrees out of phase. We can diagram the experiment as follows:

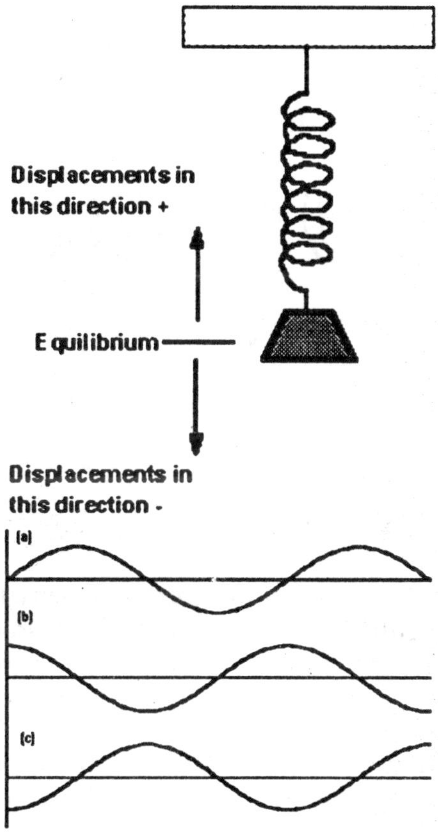

Curve (a) represents the oscillatory behaviour of the spring, curve (b) is the applied force that is required to make it oscillate and curve (c) is the applied force that is required to stop it from oscillating. In (b) the spring is absorbing energy and in (c) it is providing energy. (b) represents absorption and (c) represents stimulated emission. The only difference is the phase - 180 degrees to be exact. Both (b) and (c) are out of phase with (a).

Our description so far has applied only to a driving force in resonance with the vibrational frequency of the spring. What happens out of resonance? If the driving frequency is lower than resonance, there will be an in-phase component of the oscillation. If it is higher, there will be a 180 degree out-of-phase component. If the driving force is a cosine function, the oscillation will have a sine functionality at resonance and an additional cosine functionality at higher or lower frequencies.

A REAL ATOM OR MOLECULE

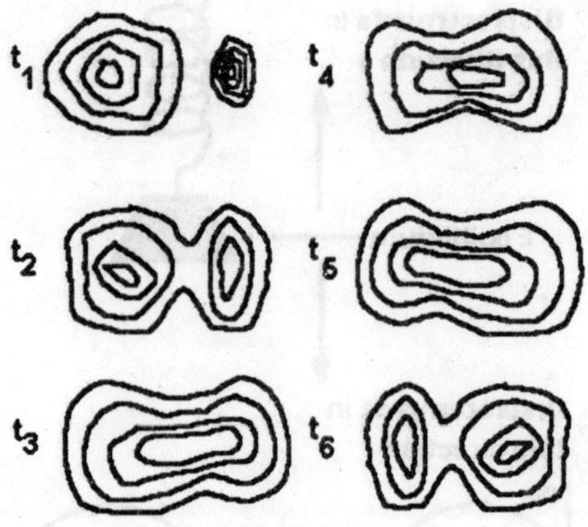

Now what happens if we apply an electric field to a real atom or molecule? The electric field will distort or polarize it. We can represent that distorted wavefunction as a linear combination of all the wavefunctions of the atom. (Actually, that isn't quite true. Only the states with opposite parity will contribute. Do you know why?) If we reverse the direction of the field, the distortion will also reverse. If we do that often at a frequency that matches a transition in the atom or molecule, the distortion or polarization will build up. An isolated atom or molecule doesn't suffer any losses and therefore the distortions can become extremely large. Quantum mechanically, one describes this problem by saying that an excited state gets mixed into the ground state by the applied electric field and that this mixing is time dependent since the field is. We could draw a picture of this oscillation for the H atom when there are an equal amount of 1s and 2p orbitals mixed into this transition state. This case corresponds to the situation when the distortion (or dipole moment) is the largest.

A series of contour plots representing the temporal behaviour will be sketched in class.

We start at t1 with a state that is primarily 1s and progress to a state at t4 that is primarly 2p and then back again. Quantitatively, we could use the time-dependent wave equation and derive a correct time dependent wavefunction for a two-state system that is interacting with an incoming electromagnetic wave whose frequency was *w*.

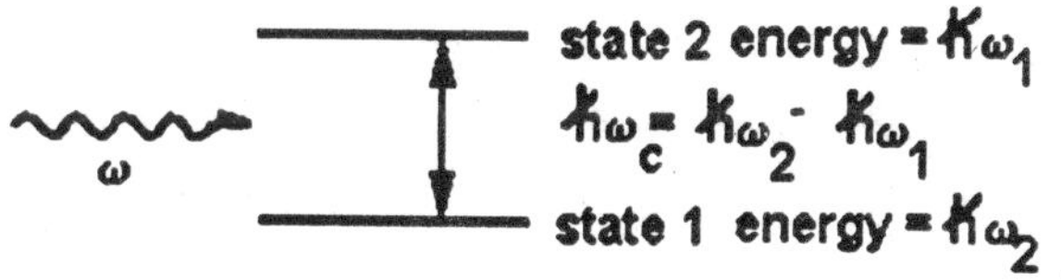

The Hamiltonian representing the interaction of the electric field, *E*, of the incoming wave with the dipole moment of the atom, u, is:

H = u *E* cos *w* t

The wavefunction as a function of time is then: (sorry too complicated for HTML, we'll do it in class)

SPECTROSCOPY

After the driving field has been applied for a long time, the effects we have been discussing will die out. They are only transients. They die out because there will always be interactions that will dephase molecules so they loose coherence with the driving fiend. Nevertheless, there will be a steandy state polarization that will have an in-phase and an out-of-phase component. If we are at resonance, we saw there was only an out-of-phase component. If we are very far from resonance, there is only an in-phase component. As an electromagnetic wave travels through a material, it induces a polarization whose oscillation launches an electromagnetic wave of its onw which adds to the incoming wave. Since there can be an out-of-phase component to this new wave, the net result is a phase shift in the emerging wave. This phase shift is the index of refraction of the material. Large phase shifts correspond to large indices of refraction. If the material is completely uniform and ordered, all of the oscillating dipoles will be related to each other by the phase and frequency of the light. The net electromagnetic field formed by the sum of all of them will add only in the direction of the propagation of the light and will destructively interfere at all other angles and all other directions. However, if there are regions where the material is not of uniform density (and in fact no material at a finite temperature can have a uniform density), then there will be some dipoles which do not contribute to the net electromagnetic field and light can be scattered in other directions. This scattering is Rayleigh scattering. If the density fluctuations are static, the scattered light will have the same frequency as the incoming light. If the density fluctuations are dynamic, the scattered frequency will be shifted by an amount that depends upon the frequency of the density fluctuations. Particular density fluctuations associated with propagating vibrations in solids (phonons) have unique frequencies and give rise to scattering at a frequency shifted from the input frequency by that of the phonon. This scattering is Brillouin scattering. Typically the shifts are in the region of 0.1 to 1 cm^{-1}.

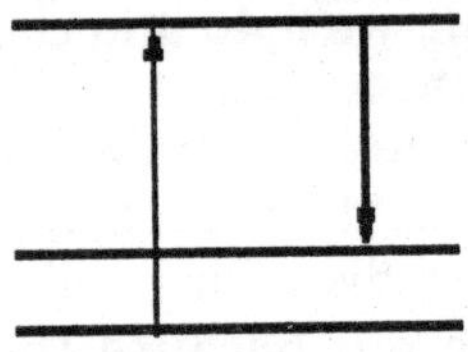

As an electromagnetic field propogates through a sample, there will be a certain average excited state population that will have a coherent and an incoherent component. At all times, the exctited state has the option of simply relaxing spontaneously and emitting a photon. This process is spontaneous emission and is incoherent relative to the driven polarization. If the spontaneous emissions occurs from the coherent excited state population, the emission is called Raman scattering. The state it relaxes to can be a different electronic state in which case it is electonic Raman scattering, a different vibrational state where it is vibrational Raman scattering (or just Raman scattering), or a different rotational state where it is rotational Raman scattering. The emitted light frequency will differ from the exciting frequency by the energy of the final state. Quantum mechanically, Raman scattering is diagramed by:

The top dotted line represents the coherent oscillating state. We know what that looks like. It is simply the atom or molecule distorted by the electric field of the incoming light wave. In a real system, it doesn't usually correspond to a particular stationary state of the molecule and since one doesn't usually want to have to derive the eigenfunctions that describe the state, we instead simply express it as a linear combination of all the states required to make this distorted state. It is called a virtual state because there is no real state at that energy usually. But we know it is simply the distorted molecule. Notice that the importance of Raman scattering depends upon how easily the molecule is distorted. Classically, this property is measured by the polarizability of the molecule and therefore polarizability determines the amount of Raman scattering.

Things become very interesting when the incoming electromagnetic wave approaches an electronic resonance. There is now the possibility of achieving a finite incoherent excited state poulation as well as a coherent population. The distortions of the molecules will become much larger near electronic resonances. Spontaneous emission will still occur from both the coherent and incoherent excited state populations which will now be considerably larger than when the driving frequency was far from resonance. The emisson from the coherent population will be resonantly enhanced Raman scattering (resonance Raman) while the incoherent population will cause normal fluorescence. There will also be losses from the excited state population due to nonradiative relaxation. All of these losses result in absorption of light.

The relative amounts of resonance Raman and fluorescence depends upon the relative coherent and incoherent excited-state populations. This in turn depends upon the relaxation effects that are important in the system. In most molecular systems, dephasing by collisions or other interactions is a dominant factor in determining the incoherent population. The ratio of fluorescence to Raman will be approximately 2 / (T2*GAMMA) where T2 is the dephasing time constant and GAMMA is the radiative rate of spontaneous emission.

The fluorescence that one observes can also have different character. If the fluorescence orginates from an incoherent population that has been formed by dephasing, emission to the ground state will be resonance fluorescence and will have a lifetime characteristic of the emitting state. The linewidth of the emission will reflect the widths of the initial and final states. In contrast, the resonance Raman emission will cease when the excitation source is turned off and will have an emission linewidth characteristic of the final state. It is also possible that the incoherent excited population can be formed from relaxation from the state that was connected to the ground state. This fluorescence is relaxed fluorescence and is generally shifted in wavlength.

Either fluorescence or resonance Raman can be quenched. If there are rapid nonradiative relaxation processes that depopulate the incoherent excited state population, the fluorescence will be quenched. One would then see only the resonance Raman. (The resonance Raman intensity could also be lowered.) If there were rapid dephasing or relaxation processes that depopulated the coherent populations, the resonance Raman emission would be quenched. For example, one can observe strong Raman scattering many times when the v=0 to v=0 electronic transition is excited but very little when the v=0 to v=n transitions are excited because the vibrational states relax very quickly to the unexcited vibrational state. Then the excitation profile of the resonance Raman effect will not follow the absorption profile for the molecule.

Chapter 11

Spectroscopy

INTRODUCTION

Spectroscopy is the use of the absorption, emission, or scattering of electromagnetic radiation by atoms or molecules (or atomic or molecular ions) to qualitatively or quantitatively study the atoms or molecules, or to study physical processes. The interaction of radiation with matter can cause redirection of the radiation and/or transitions between the energy levels of the atoms or molecules. A transition from a lower level to a higher level with transfer of energy from the radiation field to the atom or molecule is called absorption. A transition from a higher level to a lower level is called emission if energy is transfered to the radiation field, or nonradiative decay if no radiation is emitted. Redirection of light due to its interaction with matter is called scattering, and may or may not occur with transfer of energy, i.e., the scattered radiation has a slightly different or the same wavelength.

ABSORPTION

When atoms or molecules absorb light, the incoming energy excites a quantized structure to a higher energy level. The type of excitation depends on the wavelength of the light. Electrons are promoted to higher orbitals by ultraviolet or visible light, vibrations are excited by infrared light, and rotations are excited by microwaves.

An absorption spectrum is the absorption of light as a function of wavelength. The spectrum of an atom or molecule depends on its energy level structure, and absorption spectra are useful for identifying of compounds.

Measuring the concentration of an absorbing species in a sample is accomplished by applying the Beer-Lambert Law.

EMISSION

Atoms or molecules that are excited to high energy levels can decay to lower levels by emitting radiation (emission or luminescence). For atoms excited by a high-temperature energy source this light emission is commonly called atomic or optical emission (see atomic-emission spectroscopy), and for atoms excited with light it is called atomic fluorescence (see atomic-fluorescence spectroscopy) or molecular fluorescence (see molecular fluorescence spectroscopy). For molecules it is called fluorescence if the transition is between states of the same spin and phosphorescence if the transition occurs between states of different spin.

The emission intensity of an emitting substance is linearly proportional to analyte concentration at low concentrations, and is useful for quantitating emitting species.

SCATTERING

When electromagnetic radiation passes through matter, most of the radiation continues in its original direction but a small fraction is scattered in other directions. Light that is scattered at the same wavelength as the incoming light is called Rayleigh scattering. Light that is scattered in transparent solids due to vibrations (phonons) is called Brillouin scattering. Brillouin scattering is typically shifted by 0.1 to 1 cm^{-1} from the incident light. Light that is scattered due to vibrations in molecules or optical phonons in solids is called Raman scattering. Raman scattered light is shifted by as much as 4000 cm^{-1} from the incident light.

BEER-LAMBERT LAW

The Beer-Lambert law (or Beer's law) is the linear relationship between absorbance and concentration of an absorbing species. The general Beer-Lambert law is usually written as:

$$A = a(\lambda) * b * c$$

where A is the measured absorbance, $a(\lambda)$ is a wavelength-dependent absorptivity coefficient, b is the path length, and c is the analyte concentration. When working in concentration units of molarity, the Beer-Lambert law is written as:

$$A = \varepsilon * b * c$$

where ε is the wavelength-dependent molar absorptivity coefficient with units of $M^{-1}\ cm^{-1}$.

Experimental measurements are usually made in terms of transmittance (T), which is defined as:

$$T = I / I_o$$

where I is the light intensity after it passes through the sample and I_o is the initial light intensity. The relation between A and T is:

$$A = -\log T = -\log (I / I_o).$$

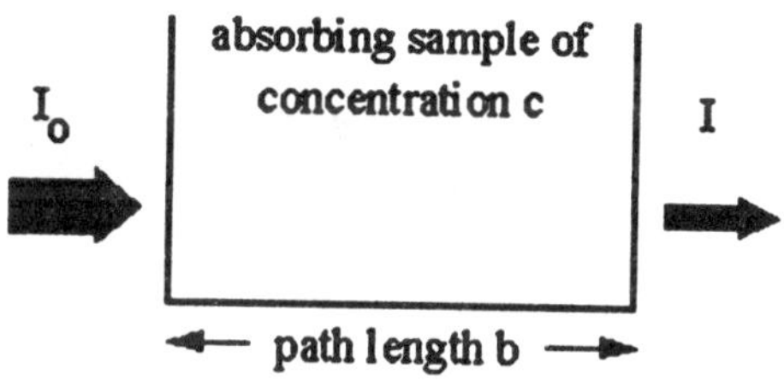

Fig. Absorption of light by a sample

Modern absorption instruments can usually display the data as either transmittance, per cent-transmittance, or absorbance. An unknown concentration of an analyte can be determined by measuring the amount of light that a sample absorbs and applying Beer's law. If the absorptivity coefficient is not known, the unknown concentration can be determined using a working curve of absorbance versus concentration derived from standards.

Derivation of the Beer-Lambert law

The Beer-Lambert law can be derived from an approximation for the absorption coefficient for a molecule by approximating the molecule by an opaque disk whose cross-sectional area,, represents the effective area seen by a photon of frequency w. If the frequency of the light is far from resonance, the area is approximately 0, and if w is close to resonance the area is a maximum. Taking an infinitesimal slab, dz, of sample:

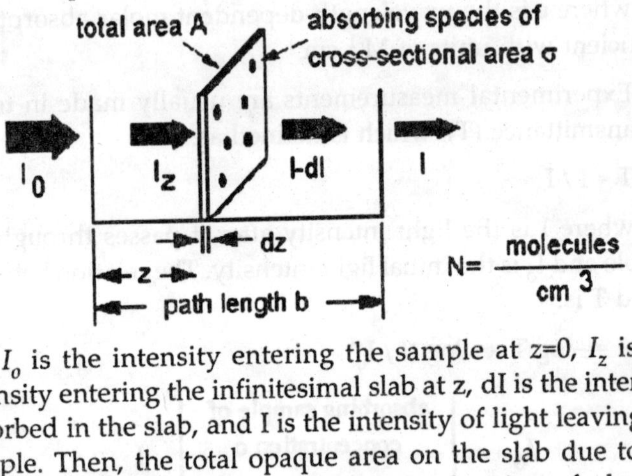

I_o is the intensity entering the sample at z=0, I_z is the intensity entering the infinitesimal slab at z, dI is the intensity absorbed in the slab, and I is the intensity of light leaving the sample. Then, the total opaque area on the slab due to the absorbers is $\sigma * N * A * dz$. Then, the fraction of photons absorbed will be $\sigma * N * A * dz$ / A so,

$$dI / I_z = - \sigma * N * dz$$

Integrating this equation from z = 0 to z = b gives:

$$\ln(I) - \ln(I_o) = - \sigma * N * b$$

$$\text{or } - \ln(I / I_o) = \sigma * N * b.$$

Since N (molecules/cm^3) * (1 mole / 6.023x10^{23} molecules) * 1000 cm^3 / liter = c (moles/liter)

and 2.303 * log(x) = ln(x)

then $- \log(I / I_o) = \sigma * (6.023\times10^{20} / 2.303) * c * b$

or $- \log(I / I_o) = A = \in * b * c$

where $\in = \sigma * (6.023x10^{20} / 2.303) = \sigma * 2.61x10^{20}$

Typical cross-sections and molar absorptivities are:

	σ (cm^2)	$\in$ (M^{-1} cm^{-1})
Absorption	atoms	10^{-12} $3x10^8$
Molecules	10^{-16}	$3x10^4$
Infrared	10^{-19}	3x10
Raman scattering	10^{-29}	$3x10^{-9}$

Limitations of the Beer-Lambert law

The linearity of the Beer-Lambert law is limited by chemical and instrumental factors. Causes of nonlinearity include:

- Deviations in absorptivity coefficients at high concentrations (>0.01M) due to electrostatic interactions between molecules in close proximity
- Scattering of light due to particulates in the sample
- Fluoresecence or phosphorescence of the sample
- Changes in refractive index at high analyte concentration
- Shifts in chemical equilibria as a function of concentration
- Non-monochromatic radiation, deviations can be minimized by using a relatively flat part of the absorption spectrum such as the maximum of an absorption band
- Stray light

QUANTITATIVE FLUORIMETRY

Light emission from atoms or molecules can be used to quantitate the amount of the emitting substance in a sample. The relationship between fluorescence intensity and analyte concentration is:

$$F = k * QE * P_o * (1-10^{[-\in *b*c]})$$

where F is the measured fluorescence intensity, k is a geometric instrumental factor, QE is the quantum efficiency (photons emitted/photons absorbed), P_o is the radiant power of the excitation source, $\in$ is the wavelength-dependent molar absorptivity coefficient, b is the path length, and c is the analyte concentration ($\in$, b, and c are the same as used in the Beer-Lambert law).

Expanding the above equation in a series and dropping higher terms gives:

$$F = k * QE * P_o * (2.303 * \in * b * c)$$

This relationship is valid at low concentrations ($<10^{-5}$ M) and shows that fluorescence intensity is linearly proportional to analyte concentration.

Determining unknown concentrations from the amount of fluorescence that a sample emits requires calibration of a fluorimeter with a standard (to determine K and QE) or by using a working curve.

LIMITATIONS

Many of the limitations of the Beer-Lambert law also affect quantitative fluorimetry. Fluorescence measurements are also susceptible to inner-filter effects. These effects include excessive absorption of the excitation radiation (pre-filter effect) and self-absorption of atomic resonance fluorescence (post-filter effect).

ATOMIC TRANSITIONS

The probability that an atomic spectroscopic transition will occur is called the transition probability or transition strength. This probability will determine the extent to which an atom will absorb light at a resonance frequency, and the intensity of the emission lines from an atomic excited state. The spectral width of a spectroscopic transition depends on the widths of the initial and final states. The width of the ground state is essentially a delta function and the width of an excited state depends on its lifetime.

Specific Documents

- Transition strengths
- Excited-state lifetime and the natural linewidth
- Transition lineshapes and broadening

SPECTROSCOPIC TRANSITION STRENGTHS

An atom or molecule can be stimulated by light to change from one energy state to another. An atom or molecule in an excited energy state can also decay spontaneously to a lower state. The probability of an atom or molecule changing states depends on the nature of the initial and final state wavefunctions, how strongly light can interact with them, and on the intensity of any incident light. This document discusses some of the practical terms used to describe the probability of a transition occuring, which is commonly called the transition strength. To a first approximation, transitions strengths are governed by selection rules which determine whether a transition is allowed or disallowed. Practical measurements of transitions strengths are usually described in terms of the Einstein A and B coefficients or the oscillator strength (f).

SELECTION RULES

1. The parity of the initial and final wavefunctions must be different.
2. The spin can not change, $\Delta S = 0$
3. The change in orbital angular momentum can be $\Delta L = 0, \pm 1$ but L=0 to L=0 transitions are not allowed.
4. The change in total angular momentum can be $\Delta J = 0, \pm 1$ but J=0 to J=0 transitions are not allowed.

TRANSITION PROBABILITY

The transition probability is R^2 with units of J cm^3, where R is the transition moment given by:

$$R = \langle X \mid u \mid X \rangle$$

and u is the dipole moment operator. Basically what this equation indicates is that the strength of a transition is relative to how strongly the dipole moment of a resonance between energy states can couple to the electric field of a light wave.

EINSTEIN COEFFICIENTS

For a two-level system (ground-state level i and upper level j), the rate of an upward stimulated transition (absorption, $-dN_i/dt$ or dN_j/dt) is:

$$\frac{-dN_i}{dt} = N_i B_{ij} U_v$$

where N_i is the number density of atoms in the ground state, U_v is the light intensity, and the proportionality factor B_{ij} is the Einstein B coefficient for absorption:

$$B_{ij} = \frac{8\pi^3 R^2}{3hg_i}$$

For stimulated emission the Einstein coefficient becomes:

$$B_{ji} = B_{ij}\frac{g_i}{g_j}$$

where g_i and g_j are the degeneracies of the ground and excited states, respectively.

Atoms in the excited state can decay without the presence of an external light field due to stimulation due to "zero-point fluctuations." Zero-point fluctuations are the dynamic variations in the shape of an electronic orbital at any instant in time. These instantaneous orbitals can be described by a linear combination of the wavefunctions of the system, which provides the mechanism for transitions between different states of the system. The spontaneous decay rate ($-dN_j/dt$ or dN_i/dt) is:

$$-dN_j/dt = N_j * A_{ji}$$

where A_{ji} is the Einstein coefficient for spontaneous emission:

$$A_{ji} = \frac{8\pi h B_{ij} g_i}{g_j \pi_m^3} = \frac{64\pi^4 R^2}{3hg_j \pi_m^3}$$

Since atoms in the upper level can decay by both spontaneous and stimulated emission, the total downward rate ($-dN_j/dt$ or dN_i/dt) is given by:

$$\frac{-dN_j}{dt} = N_j \left(A_{ji} + B_{ji} U_v\right)$$

OSCILLATOR STRENGTH

The oscillator strength of a transition is a dimensionless number that is useful for comparing different transitions. It is defined as the ratio of the strength an atomic or molecular transition to the theoretical transition strength of a single electron using a harmonic-oscillator model. For absorption:

$$F_{ij} = \frac{4 \epsilon_0 \, hcm_e B_{ij}}{e^2 \lambda_m}$$

and for emission:

$$f_{ji} = f_{ij}\, g_i/g_j$$

Oscillator strengths can range from 0 to 1, or a small integer. A strong transition will have an f close to 1. Oscillator strengths greater than 1 result from the degeneracy of real electronic systems.

Tabulations in the literature often use gf, where

$$gf = g_i f_{ij} = g_j f_{ji}$$

EXCITED-STATE LIFETIME AND NATURAL LINEWIDTH

The excited state of an atom will have an intrinsic lifetime due to radiative decay given by:

$$-dN_j/dt = N_j \, \Sigma_{(i<j)} A_{ji}$$

where the A_{ji}s are the Einstein spontaneous emission coefficients for all of the radiative transitions originating from level j. Intergrating this equation produces:

$N_j(t) = N_j(0)\, e^{(-t/\tau_j)}$

where τ_j is defined as the radiative lifetime:

$1 / \Sigma_{(i<j)} A_{ji}$

Strong atomic transitions have A_{ji}s of 10^8 to 10^9 s^{-1}, so lifetimes are 1 to 10 ns. Lifetimes can be shortened by collisions or stimulated emission.

NATURAL LINEWIDTH OF ATOMIC TRANSITIONS

The natural linewidth (the intrinsic linewidth in the absence of external influences) of an energy level is determined by the lifetime due to the Heisenberg uncertainty principle:

$\Delta E^*\Delta t \cong h / 2\pi$

So the natural width of an energy level is:

$$\Delta E_j = \frac{h}{2\pi\tau_j}$$

or

$$\Delta E_j = \frac{h\sum_{(i,j)A_{ji}}}{2\pi\tau_j}$$

Since $E = h\upsilon$

$$\Delta\upsilon = \sum_{(i<j)A_{ji}}$$

where $\Delta\upsilon$ is the linewidth in frequency units of a transition between an excited state and the ground state. Since the ground state has an essentially infinite lifetime, the transition linewidth is governed by the width of the excited state.

The lineshape of a transition with only natural broadening is a Lorentzian.

SPECTROSCOPIC LINESHAPES

The width and shape of spectroscopic transitions will affect the ability to extract qualitative and quantitative information from a spectrum. The lineshapes of spectroscopic transitions depend on the broadening mechanisms of the initial and final states, and include natural broadening, collisional broadening, power broadening, and Doppler broadening. Natural, collisional, and power broadening are homogeneous mechanisms and produce Lorentzian lineshapes, and Doppler broadening is a form of inhomogeneous broadening and has a Gaussian lineshape. The general form of Lorentzian and Gaussian lineshapes are shown below. Combinations of Lorentzian and Gaussian lineshapes can be approximated by a Voigt profile.

Lorentzian (Upper) and Gaussian (lower) lineshapes normalized to equal maximum

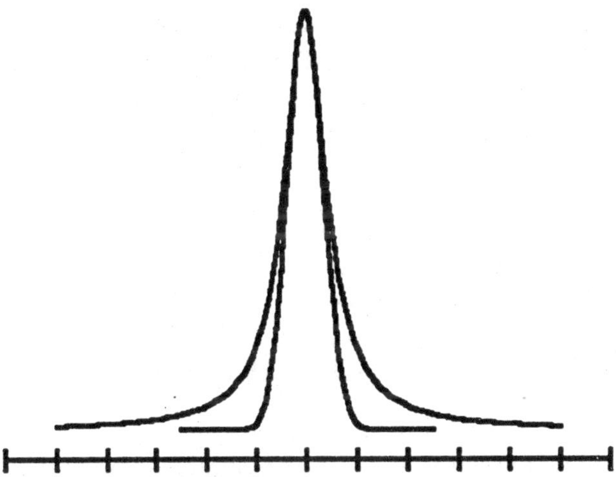

COLLISIONAL OR PRESSURE BROADENING

Collisions broaden spectroscopic linewidths by shortening the lifetime of the excited states.

Power Broadening

Power broadening occurs by shortening the lifetime of the excited state due to stimulated emission.

Doppler Broadening

Doppler broadening is due to the distribution of atomic velocities (speed and direction), which each have a Doppler shift with respect to an observer.

ATOMIC-ABSORPTION SPECTROSCOPY (AA)

Atomic-absorption (AA) spectroscopy uses the absorption of light to measure the concentration of gas-phase atoms. Since samples are usually liquids or solids, the analyte atoms or ions must be vapourized in a flame or graphite furnace. The atoms absorb ultraviolet or visible light and make transitions to higher electronic energy levels. The analyte concentration is determined from the amount of absorption. Applying the Beer-Lambert law directly in AA spectroscopy is difficult due to variations in the atomization efficiency from the sample matrix, and nonuniformity of concentration and path length of analyte atoms (in graphite furnace AA). Concentration measurements are usually determined from a working curve after calibrating the instrument with standards of known concentration.

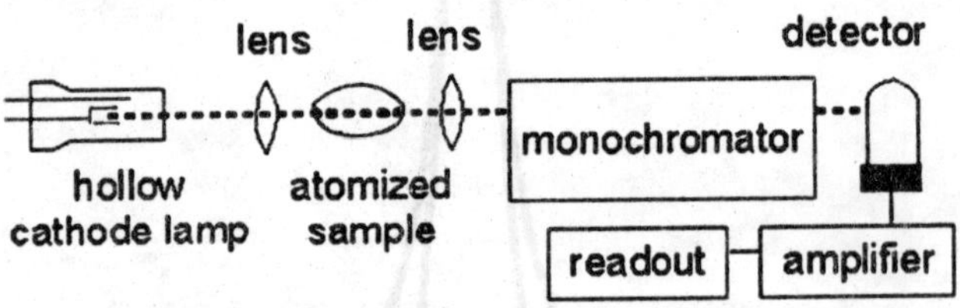

Fig. Schematic of an atomic-absorption experiment

LIGHT SOURCE

The light source is usually a hollow-cathode lamp of the element that is being measured. Lasers are also used in research instruments. Since lasers are intense enough to excite atoms to higher energy levels, they allow AA and atomic fluorescence measurements in a single instrument. The disadvantage of these narrow-band light sources is that only one element is measurable at a time.

Atomizer

AA spectroscopy requires that the analyte atoms be in the gas phase. Ions or atoms in a sample must undergo desolvation and vapourization in a high-temperature source such as a flame or graphite furnace. Flame AA can only analyse solutions, while graphite furnace AA can accept solutions, slurries, or solid samples.

Flame AA uses a slot type burner to increase the path length, and therefore to increase the total absorbance. Sample solutions are usually aspirated with the gas flow into a nebulizing/mixing chamber to form small droplets before entering the flame.

The graphite furnace has several advantages over a flame. It is a much more efficient atomizer than a flame and it can directly accept very small absolute quantities of sample. It also provides a reducing environment for easily oxidized elements. Samples are placed directly in the graphite furnace and the furnace is electrically heated in several steps to dry the sample, ash organic matter, and vapourize the analyte atoms.

LIGHT SEPARATION AND DETECTION

AA spectrometers use monochromators and detectors for uv and visible light. The main purpose of the monochromator is to isolate the absorption line from background light due to interferences. Simple dedicated AA instruments often replace the monochromator with a bandpass interference filter. Photomultiplier tubes are the most common detectors for AA spectroscopy.

Fig. Picture of a flame atomic-absorption spectrometer

ATOMIC EMISSION SPECTROSCOPY (AES, OES)

Atomic emission spectroscopy (AES or OES [optical emission spectroscopy]) uses quantitative measurement of the optical emission from excited atoms to determine analyte concentration. Analyte atoms in solution are aspirated into the excitation region where they are desolvated, vapourized, and atomized by a flame, discharge, or plasma. These high-temperature atomization sources provide sufficient energy to promote the atoms into high energy levels. The atoms decay back to lower levels by emitting light. Since the transitions are between distinct atomic energy levels, the emission lines in the spectra are narrow. The spectra of samples containing many elements can be very congested, and spectral separation of nearby atomic transitions requires a high-resolution spectrometer. Since all atoms in a sample are excited simultaneously, they can be detected simultaneously using a polychromator with multiple detectors. This ability to simultaneously measure multiple elements is a major advantage of AES compared to atomic-absorption (AA) spectroscopy.

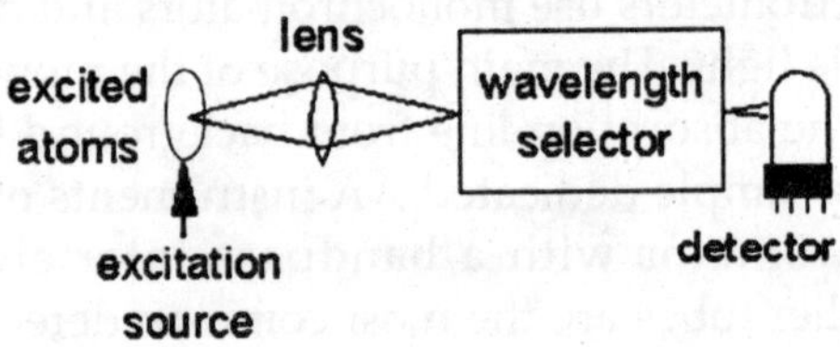

Fig. Schematic of an AES experiment

As in AA spectroscopy, the sample must be converted to free atoms, usually in a high-temperature excitation source. Liquid samples are nebulized and carried into the excitation source by a flowing gas. Solid samples can be introduced into the source by a slurry or by laser ablation of the solid sample in a gas stream. Solids can also be directly vapourized and excited by a spark between electrodes or by a laser pulse. The excitation source must desolvate, atomize, and excite the analyte atoms. A variety of excitation sources are described in separate documents:

- Direct-current plasma (DCP)
- Flame
- Inductively-coupled plasma (ICP)

- Laser-induced breakdown (LIBS)
- Laser-induced plasma
- Microwave-induced plasma (MIP)
- Spark or arc

DIRECT-CURRENT PLASMA EXCITATION SOURCE

A direct-current plasma (DCP) is created by an electrical discharge between two electrodes. A plasma support gas is necessary, and Ar is common. Samples can be deposited on one of the electrodes, or if conducting can make up one electrode. Insulating solid samples are placed near the discharge so that ionized gas atoms sputter the sample into the gas phase where the analyte atoms are excited. This sputtering process is often referred to as glow-discharge excitation.

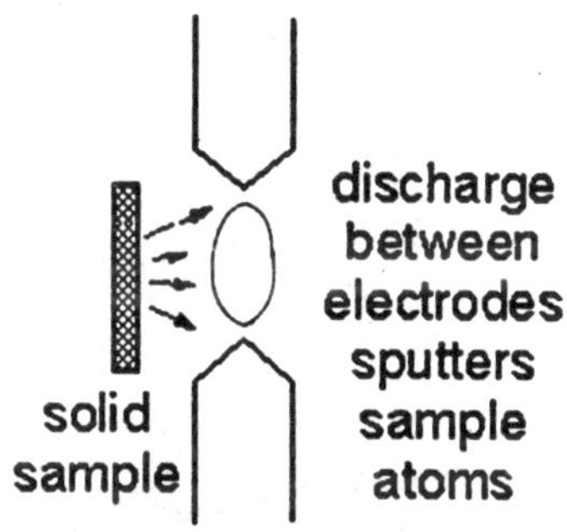

FLAME EXCITATION SOURCE

A flame provides a high-temperature source for desolvating and vapourizing a sample to obtain free atoms for spectroscopic analysis. In atomic absorption spectroscopy ground state atoms are desired. For atomic emission spectroscopy the flame must also excite the atoms to higher energy levels. The table lists temperatures that can be achieved in some commonly used flames.

Table: Temperatures of some common flames

Fuel	Oxidant	Temperature (K)
H_2	Air	2000-2100
C_2H_2	Air	2100-2400

H_2	O_2	2600-2700
C_2H_2	N_2O	2600-2800

The figure shows a total consumption burner in which the sample solution is directly aspirated into the flame. This flame design is common for atomic emissions pectroscopy.All desolvation, atomization, and excitation occurs in the flame. Other flame designs nebulize the sample and premix it with the fuel and oxidant before it reaches the burner. Atomic-absorption instruments almost always use a nebulizer and also use a slot burner to increase the path length for the sample absorption.

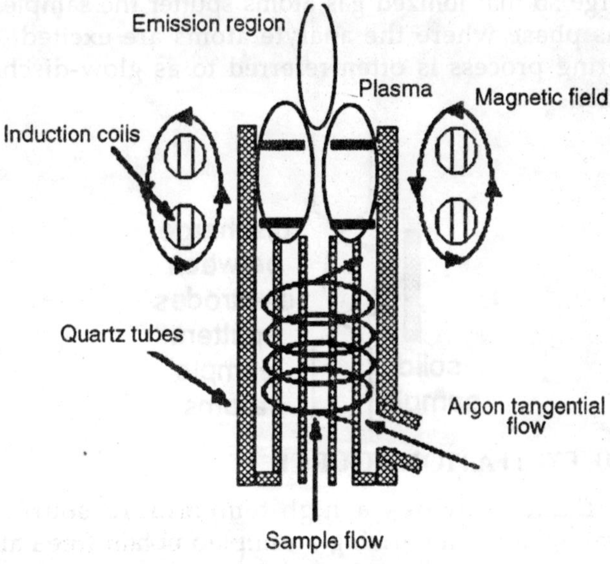

INDUCTIVELY-COUPLED PLASMA (ICP) EXCITATION SOURCE

An inductively coupled plasma (ICP) is a very high temperture (7000-8000K) excitation source that efficiently desolvates, vapourizes, excites, and ionizes atoms. Molecular interferences are greatly reduced with this excitation source but are not eliminated completely. ICP sources are used to excite atoms for atomic-emission spectroscopy and to ionize atoms for mass spectrometry.

Instrumentation

The sample is nebulized and entrained in the flow of plasma support gas, which is typically Ar. The plasma torch consists of concentric quartz tubes.

The inner tube contains the sample aerosol and Ar support gas and the outer tube contains flowing gas to keep the tubes cool. A radiofrequency (RF) generator (typically 1-5 kW @ 27 MHz) produces an oscillating current in an induction coil that wraps around the tubes. The induction coil creates an oscillating magnetic field, which produces an oscillating magnetic field The magnetic field in turn sets up an oscillating current in the ions and electrons of the support gas (argon).

As the ions and electrons collide with other atoms in the support gas Since the atomic emission lines are very narrow, a high-resolution polychromator is needed to selectively monitor each emission line.

LASER-INDUCED BREAKDOWN EXCITATION SOURCE

When a high-energy laser pulse is focused into a gas or liquid, or onto a solid surface, it can cause dielectric breakdown and create a hot plasma. For solids the laser pulse also ablates material into the gas phase. The energy of the laser-created plasma can atomize, excite, and ionize analyte species, which can then be detected and quantified by atomic-emission spectroscopy or mass spectrometry.

LASER-INDUCED PLASMA EXCITATION SOURCE

A high-power CO_2 laser that is focused into a support gas, such as Ar, can maintain a hot plasma. The energy of the plasma can atomize, excite, and ionize analyte species present in the support gas, which can then be detected and quantified by atomic-emission spectroscopy or mass spectrometry. It can also be used in a glow-discharge mode to sputter analyte atoms off of a solid surface for analysis in the plasma.

MICROWAVE-INDUCED PLASMA EXCITATION SOURCE

A microwave-induced plama consists of a quartz tube surrounded by a microwave waveguide or cavity. Microwaves produced from a magnetron (a microwave generator) fill the waveguide or cavity and cause the electrons in the plasma support gas to oscillate. The oscillating electons collide with other atoms in the flowing gas to create and maintain a high-temperature plasma. As in inductively coupled plasmas, a spark is needed to create some initial electrons to create the plasma. Atomic emission is measured from excited analyte atoms as they exit the microwave waveguide or cavity.

SPARK AND ARC EMISSION SOURCES

Spark and arc excitation sources use a current pulse (spark) or a continuous electical discharge (arc) between two electrodes to vapourize and excite analyte atoms. The electrodes are either metal or graphite. If the sample to be analysed is a metal, it can be used as one electrode. Non-conducting samples are ground with graphite powder and placed into a cup-shaped lower electrode. Arc and spark sources can be used to excite atoms for atomic-emission spectroscopy or to ionize atoms for mass spectrometry. Arc and spark excitation sources have been replaced in many applications with plasma or laser sources, but are still widely used in the metals industry.

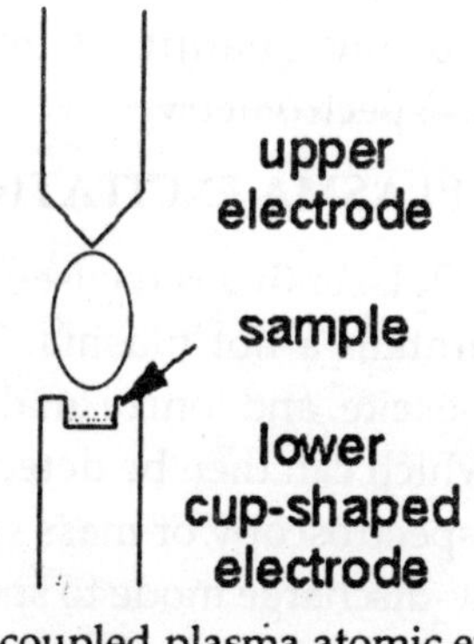

Fig. An inductively-coupled plasma atomic emission spectrometer

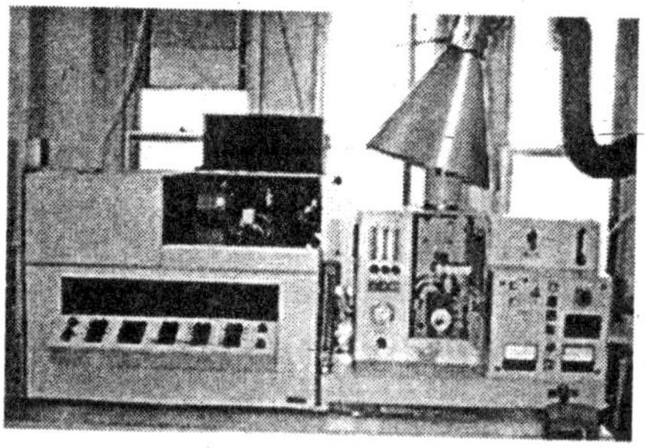

Fig. Optical Spectrometers

OPTICAL SPECTROMETERS

A spectrometer is an optical system that transmits a specific band of electromagnetic spectrum. Dispersion of different wavelengths is accomplished with the separating capability of refraction (prism) or diffraction (diffraction grating). Typical applications are isolation of a narrow band of radiation from a continuum light source for absorption measurements, or analysis of the emission from excited atoms or molecules.

MONOCHROMATOR DESIGNS

A typical monochromator design is shown below. It consists of the diffraction grating (dispersing element), slits, and spherical mirrors.

Schematic of a Czerny-Turner monochromator

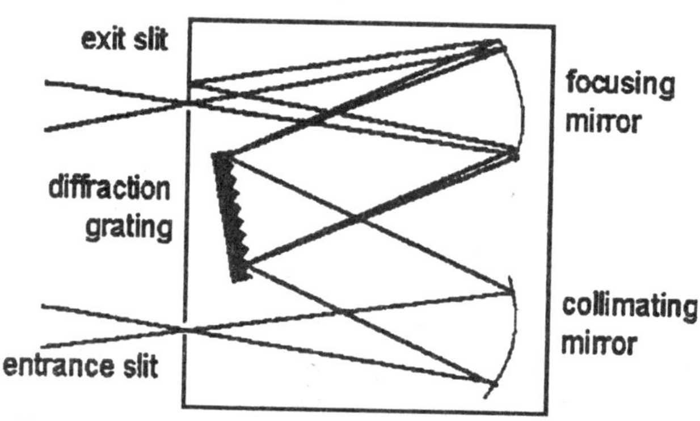

Fig. Scanning is accomplished by rotating the grating.

MONOCHROMATOR PARAMETERS

Bandpass

The wavelength range that the monochromator transmits.

Dispersion

The wavelength dispersing power, usually given as spectral range / slit width (nm/mm). Dispersion depends on the focal length, grating resolving power, and the grating order.

Resolution

The minimum bandpass of the spectrometer, usually determined by the aberrations of the optical system.

Acceptance angle (f/#)

A measure of light collecting ability, focal length / mirror diameter

Blaze wavelength

The wavelength of maximum intensity in first order.

Spectrograph

A spectrometer that records a wide bandpass with a photographic plate or an array detector. The spectrometer requires a flat image field.

Polychromator

A spectrometer with multiple detectors for simultaneous detection of multiple analytes.

ATOMIC-FLUORESCENCE SPECTROSCOPY (AFS)

Atomic fluorescence is the optical emission from gas-phase atoms that have been excited to higher energy levels by absorption of electromagnetic radiation. The main advantage of fluorescence detection compared to absorption measurements is the greater sensitivity achievable because the fluorescence signal has a very low background. The resonant excitation provides selective excitation of the analyte to avoid interferences. AFS is useful to study the electronic structure of atoms and to make quantitative measurements. Analytical applications include flames and plasmas diagnostics, and enhanced sensitivity in atomic analysis. Because of the differences in the nature of the energy-level structure between atoms and molecules, discussion of laser-induced fluorescence (LIF) from molecules is found in a separate document.

Analysis of solutions or solids requires that the analyte atoms be desolvated, vapourized, and atomized at a relatively low temperature in a heat pipe, flame, or graphite furnace. A hollow-cathode lamp or laser provides the resonant excitation to promote the atoms to higher energy levels. The atomic fluorescence is dispersed and detected by monochromators and photomultiplier tubes, similar to atomic-emission spectroscopy instrumentation.